Schools Where Children Matter

Exploring Educational Alternatives

Doralice Lange DeSouza Rocha

P.O. Box 328, Brandon, VT 05733-0328.
1-800-639-4122
www.great-ideas.org

Acknowledgments

First of all, I would like to thank the teachers who participated in my research project and who so openly let me enter their classrooms and lives. I really appreciate their willingness and courage to participate in this study knowing that their thoughts and work—perhaps not only their best, but also their worst—would become public.

Second, I would like to thank the Conselho Nacional de Pesquisa e Desenvolvimento Tecnológico (CNPq) (the CNPq is an office of the Brazilian government whose goal is to foster the development of science and technology in Brazil), the Foundation for Educational Renewal, and the Universidade Tuiuti do Paraná (UTP) for the financial support that made this research and book project possible. The CNPq financed my doctoral studies at the Harvard Graduate School of Education where I developed one of the case studies I present here. Both the Foundation for Educational Renewal and the UTP sponsored the remaining research. The Foundation is also publishing this work.

Third, I would like to thank Ron Miller for supporting my research from the beginning to the end, from giving me feedback on my initial research project to helping me edit the final drafts. I am indebted to the late Donald Oliver as well as to Ba and Josette Luvmor and Scott Forbes for helping me see some fundamental issues in the development of my research which I would perhaps not have been able to see without their help. And finally, I would like to thank Claudia Berman for sharing insights on my research on the School Around Us, and David Marshak and Karen Weilland for giving me feedback on the research on Waldorf Education.

Finally, I would like to thank my family for their patience and support. Without the help of my mother, grandmother, and husband mothering my children and taking care of me, I would not have been able to dedicate my time and energy to this work. I also would like to thank my children, Isadora (5 years old) and Cauã (6 months old) who unknowingly have inspired and given me strength to complete this project. I hope this book can somehow add to the efforts of others to enrich the quality of education that children receive so that they can develop to their fullest potential and become the best they are meant to be.

Table of Contents

The Bellwether School

Foreword

In today's educational and political climate, families enjoy more choices among different types of schools, and different styles of teaching and learning, than ever before. The proliferation of charter schools, the growing popularity of alternative methods such as Montessori and Waldorf education in both public and independent schools, the growth of the homeschooling movement, and new initiatives to provide public funding to families choosing private education are giving parents an expanding range of options. The question is, how should parents decide which of these choices are best for their children? How do they know which educational approach best reflects their own family goals or values, or their children's personal ways of learning?

It is extremely unfortunate that the opening up of educational choices has been accompanied by a narrowing down of educational purposes. The policies that break down inflexible structures of schooling are yet imposing an equally inflexible understanding of what teaching and learning should look like or aim for. In short, despite the expansion of school choice, education is being increasingly standardized, and it is simply assumed that the "best" schools are those that crank out the highest scores on objective tests. Education is increasingly concerned with the memorization of the "facts" authorized by state bureaucracies and enshrined in textbooks. Parents can choose schools, but they are under significant pressure to choose those that are most efficient in drumming officially approved information into their children's minds. When you think about where this could be leading us, this is rather frightening.

The opening of new educational options could, instead, be an opportunity for parents and children to discover excitingly different kinds of teaching and learning! For well over a century now, bold educational pioneers have demonstrated that it is entirely possible to provide learning environments that nourish a wide range of young people's latent energies: their creativity and resourcefulness, their sociability and caring, their problem-solving and critical inquiry, above all their authentic enthusiasm for learning. *Education can aim for more than mindless memorization of arbitrary "facts"*. It can be *holistic* in the sense of cultivating all the capacities that make us human. These educational visionaries—Francis Parker and John Dewey, Rudolf Steiner and Maria Montessori, A. S. Neill and John Holt, Krishnamurti and Loris Malaguzzi, to name some of the most influential—have shown us how standardization and obsessive measurement impoverish students' learning, because young people taught according to their methods grow into maturity with more joy, confidence, resourcefulness and self-understanding. They are more complete and fulfilled human beings. They retain their individuality yet participate in community with enthusiasm and purpose.

In this book, Doralice Lange de Souza Rocha shows us just why this is so. Her careful, thoughtful observations of three "holistic" schools reveal that children thrive when they are treated as valued individuals rather than as standardized units. They blossom when the deeper dimensions of their personalities are respected and given room to explore and to grow. The three schools profiled here are quite different from each other, demonstrating that holistic education is not a rigid ideology but a flexible attitude that accommodates the needs and goals of different communities, different teachers, different learners. This is precisely why holistic education is different from, and superior to, standardized education. People matter more.

I think the most important point Dr. Rocha makes in this book is that *education should follow the needs of the learner.* Comparing these different schools, she concludes that no one of them is best for all children. Of course, there is an internal consistency within the logic of each approach. What I call "libertarian" education (as practiced by Kendall, the teacher at the School Around Us described in this book) makes perfect sense when its major premise, that children will naturally explore the world and learn significantly from it when given responsibility for their own learning, matches a specific child's inclination and tendencies. From this perspective, the Waldorf approach appears mystifyingly rigid, even oppressive. Yet Rudolf Steiner's philosophy has its own internal consistency, and works wondrously for children of a certain temperament. While surely there is a universal spiritual core permeating all human life, our complex diversity is nothing short of amazing and lends itself to numerous educational approaches, just as it produces numerous cultural expressions, religious beliefs, artistic tastes and political ideologies.

The question is no longer "what school (or educational method) is 'best' in an absolute sense?" but "what does this child, this unique and precious being in my care (as a parent or teacher) need to develop his or her own distinct potentials?" Standardized education cannot answer that question, and indeed never bothers to ask it. Holistic educators, as we see in this heartwarming book, never stop asking it.

I commissioned Dr. Rocha to write this book because I wanted the stories of holistic schools to be told. I have written a great deal, myself, about the philosophy of holistic education and its historical and cultural sources, but this sort of writing tends to appeal to a fairly limited, academic audience. It's time for parents to discover that there are meaningful alternatives to standardized learning and relentless testing. I asked the author to include the school that I helped organize (the Bellwether School) not to make me feel important, because I have had little to do with the actual running of the school, but to bring public recognition to the beautiful way that the teachers in that school engage with children. Just as I am struck every day when I walk into the school, I wonder how any parent or teacher can remain content with the learning factories our society provides for our children when they see that something more deeply real and authentic is possible.

For a brief period in the 1960s and early 1970s, a group of passionate educators raised similar questions about schooling in modern society. Several books, such as Neill's *Summerhill*, Holt's *How Children Fail*, and George Dennison's *The Lives of Children*, among others, were widely read and helped spark an educational revolution that saw the rise of hundreds of independent "free schools," substantive innovation in public education (including schools of choice), and the birth of the modern homeschooling movement. The revolution was quickly put down by the political and cultural forces that have given us standards and high stakes testing, but the questions remain unanswered and the critique remains valid. It is time to revisit them, and Dr. Rocha takes up this challenge. She encourages readers to explore the unfamiliar territory of holistic education more deeply, and I second this invitation. There is an exciting and growing literature available to you that will show how we might begin to practice more humane, democratic, and life-affirming ways of educating our children. I hope you will join us.

Dr. Ron Miller, President
Foundation for Educational Renewal

Introduction

Sometimes we hear comments such as "Oh yes, I do teach in a successful school," or "My child does study in a wonderful school." But what exactly does a successful or wonderful school mean to us? Does it mean that it leads students to prestigious colleges and well-paying and high-status jobs? For some people, that is it. But from my perspective, this kind of outcome is not enough to qualify an educational institution as wonderful or successful. Many schools, while they promote these goals, also foster the development of fragmented, detached, and disenchanted human beings. These are people who have a poor inner life, who cannot find meaning for their existence, who cannot consistently enjoy the beauty of the world around them, and who have a permanent sense of restlessness, since they feel empty inside and nothing seems to be enough to satisfy them. Successful schools, I believe, go beyond just promoting intellectual development. They foster the development of the whole person (all aspects of the person) and address the person as a whole (as a part of a larger system that includes the family, the community he/she lives in, the natural environment, and the whole universe around him/her).

There is an alternative trend in education, named by many authors as *holistic education*, that takes into consideration the development of the whole child and addresses the child as a whole being. One of the main beliefs of the holistic education movement is that educational experiences should promote a more balanced development of—and cultivate the relationships among—all aspects of the human organism (intellectual, physical, spiritual, emotional, and social). Another main belief of the holistic education movement is that educational processes should cultivate the connections between the learner and what is to be learned, between reason and emotion, between the individual and other people, between the individual and the natural environment, among the various fields of knowledge, and among different forms of knowing. According to holistic educators, all learning should be experiential. In other words, children should be active participants in their learning processes. Educational experiences should draw forth what is already within each child, and promote a worldview less focused on materialistic concerns and more centered on the development of the full human potential and on the quality of human relationships.[1] The concept of holistic education has been described in detail by Scott Forbes (1999), Ramon Gallegos (1999, 2000), Yoshiharu Nakagawa (2000), John P. Miller (1988), and Ron Miller (1997, 2000), and I have explained it in an earlier study (de Souza, 2000).

Holistic education, as it is defined here, and as it was stated in the first issue of *Holistic Education Review*[2] in 1988 is *not* a curriculum package. It is a body of literature that has been inspiring many different kinds of educators. The holistic education movement houses different educational approaches and kinds of schools. The goal of this book is to present the findings of a long-term research project that involves case studies of three schools that define themselves as holistic. These case studies are meant to offer us thick and contextualized descriptions of some possible ways of thinking about and of translating holistic education ideas into practice at the elementary school level, and to explore some complexities involved in the implementation of holistic ideas in the real world of schools.

Background Information on the Research

When I started my doctoral studies at the Harvard Graduate School of Education in 1994, I took a class with my advisor Donald Oliver and it was then that I had my first contact with the literature on holistic education. As I read this literature, I realized that much of what was said there, even thought it was still disorganized, was exactly what I was trying to articulate in my own work and did not know how to express with words. Therefore, I decided to develop a literature review on the subject of holistic education for my qualifying paper. This literature review not only helped me understand the concept of holistic education, but revealed that there was a major gap in the field. There was too little written about the practice of this kind of education in the real world in schools, and a lack of independent scholarly research demonstrating how schools that see themselves as holistic actually implement holistic education ideas in their daily practices. This led me to undertake an empirical study on holistic education for my dissertation, an in-depth case study of a school that considered itself holistic and that actually implemented some of the main themes of the holistic education movement (de Souza, 2000).

The site at which I chose to develop my dissertation research was the School Around Us, a little school located in Maine. I investigated the educational philosophy and practice of this school and discussed the main difficulties it faced as it tried to implement its goals. One of the main findings of my study was that the primary challenge facing this school was to determine an appropriate degree of freedom and structure in the curriculum. While some members of the school community believed that children should have almost complete freedom, others thought children's school experiences should be more structured in order to promote the holistic development of children. While some thought that too much guidance on the part of the teachers would inevitably impair the children's ability to think for themselves, to make decisions on their own, and to live their own lives, others thought that too much freedom could limit children's ability to be truly free when they grew older, since then they would perhaps lack some of the tools they needed (e.g., academic knowledge and skills and self-discipline) to achieve their dreams.

The different viewpoints of the members of the school on this issue and my own lack of clarity on this matter motivated me to further explore this subject. So I decided to investigate the educational philosophy and practice of two other schools that defined themselves as holistic and that had contrasting views on the issue of freedom and structure, a Waldorf School, which I will call here "Sunshine Waldorf School," and an alternative school in Vermont called the Bellwether School. This book is meant to present the findings of my research both on the School Around Us and on these two other schools. I expect that this work will contribute different examples of how holistic education ideas can be implemented in the real world and show how different schools may deal with the issue of freedom and structure while trying to be holistic and to promote the development of the whole child.

This book was written for teachers, parents, and the public in general interested in educational issues. While I speak here as a researcher, at times I also bring in my voice as a teacher and/or as a parent. I have divided the book into four sections. The first three can be read separately and independently, since they are each case studies of the schools included in my research. In the first section, I will be talking about the school I studied for my dissertation, the School Around Us, whose main characteristic is that children have a large degree of freedom and determine, to a large extent, their own curriculum. In the second section, I will present the case of the Sunshine Waldorf School, a traditional Waldorf School, whose main characteristic is that the curriculum is highly structured and children are guided almost step by step in almost everything they do at school. In the third section, I will talk about the Bellwether School, a place where the children have a degree of freedom to make certain decisions about their own education, but within a well-established framework. In the last section, I will make some comparisons between the schools—their most evident similarities and differences—and make recommendations for future research in the field of holistic education.

Please see Appendix I for details on the research methodology utilized for this work. For now I only would like to note that, while I did preserve the real names of two of the schools—the School Around Us and the Bellwether School—and the names of the teachers who work there, I use a fictitious name for the Waldorf School and the teachers at this institution since they did not give me consent to use their real names. I changed the names of all children and parents from the three schools.

Some Introductory Words on the Issue of Freedom and Structure in Education

As we all know, there are different perspectives on what freedom means. In the Western world, for instance, when we talk about freedom, we are usually referring to absence of authority, restraints, influences, and obstructions. We also tend to relate freedom to the concepts of democracy, self-determination, autonomy, self-dependence, and self-regulation. Besides, we think of it as the having of power to do what we want to do and an ability to satisfy our desires and give free play to our impulses and expression. In the Eastern world, on the other hand, the idea of freedom doesn't necessarily relate to external factors such as these. It relates mainly to an inner condition of the mind. It usually refers to a state of enlightenment which involves an absence of conditionings, fears, or attachments of any sort.

Considering that the idea of "freedom and structure" is an important theme of this book, I would like to make here a brief introduction of different perspectives on this issue by authors who have somehow (directly or indirectly) contributed to the development of the literature on holistic education.

One important perspective, which is embraced by some holistic educators, is supported by authors such as A.S. Neill (1960, 1966), Daniel Greenberg & Mimsy Sadofsky (1992), and John Holt (1972,1982,1983). The advocates of this view have a deep trust in human nature. They believe, for example, that we are inherently good and that if we offer children a rich learning environment and support them in their pursuits, they can and will educate themselves. From their perspective, the role of the teacher is not to direct students, but to assist them as they ask for help. Also from their perspective, anything children say they are interested in learning is important. In other words, for them, academic experiences are not more significant than other kinds of experiences children might want to have. For them, freedom means having few restraints and plenty of choices. Holt, for instance, used to say that if there are to be rules in schools, that these rules be about what should not be done, instead of what must be done, so that children can feel free to explore their interests. Neill insisted that we should restrict children's actions only when strictly necessary to ensure their safety (e.g., to prevent a two-year-old from playing in the middle of a busy street) and to prevent children from interfering with other people's freedom. In his Summerhill School, students could—and still can—go or stay away from lessons as they please. However, they cannot play the drums in the middle of the night if others demand silence to sleep.

The Swiss romantic philosopher Jean Jacques Rousseau (1957) was also a firm believer in the goodness and wisdom of human nature. From his viewpoint, we as educators should never impose on the child our own knowledge and values. He believed that we should deeply respect the natural impulses of the child and leave her free to make choices and explore her personal interests. But, unlike Neill and his followers who were in favor of giving children almost complete freedom, Rousseau defended the idea that we should covertly plan learning situations for them so that they can always make good choices about their own education and life. He explained:

> Let him [the child] always believe he is the master, and let it always be you who are. There is not subjection so perfect as that which keeps the appearance of freedom.... Doubtless he ought to do only what he wants; but he ought to want only what you want him to do. He ought not to make a step without your having foreseen it; he ought not to open his mouth without your knowing what he is going to say.... Thus, not seeing you eager to oppose him, not distrusting you, with nothing to hide from you, he will not deceive you, he will not lie to you, he will fiercely show himself precisely as he is. You will be able to study him at your complete ease and arrange all around him the lessons you want to give him without his ever thinking he is receiving any." (Rousseau, 1957, p. 120)

From Rousseau's perspective, while the child should always think of herself as making decisions about her own education, we as teachers should subtly lead her to situations likely to foster her growth and development. He believed that, as the child thinks that she is in charge of her education, and as she practices making decisions about her own experiences, she learns to think for herself and progressively becomes self-sufficient.

John Dewey is another author, often quoted by advocates of holistic education, who directly deals with the issue of freedom and structure. For Dewey (1964), while the idea of freedom does involve choice, it also necessarily involves intelligence, knowledge, and certain objective conditions. It involves intelligence and knowledge, since intelligence and knowledge determine our abilities to thoroughly observe, to foresee the consequences of our choices and actions, to make plans and judgements, to be effective in what we do, and to change the course of our plans and actions as needed. It involves objective conditions because concrete conditions determine the extent to which we are able to think and carry out plans.

Freedom for Dewey also involves self-control, which from his perspective, can only be attained progressively as the child develops and matures. But note that for him, while self-control should be one of the main goals of education, teachers should carefully guide students until they can make intelligent decisions on their own, otherwise, they will not be able to be really free.

> The mere removal of external control is no guarantee for the production of self-control.... Impulses and desires that are not ordered by intelligence are under the control of accidental circumstances. It may be a loss rather than a gain to escape from the control of another person only to find one's conduct dictated by immediate whim and caprice; that is, at the mercy of impulses into whose formation intelligent judgment has not entered. A person whose conduct is controlled in this way has at most only the illusion of freedom. Actually he is directed by forces over which he has no command. (Dewey, 1963, p. 64)

Another contribution by Dewey (1963) that relates to the issue of freedom and structure derives from his theory of experience. According to him, learning occurs as new experiences build upon experiences we have already had. The richer our previous experiences, the richer our possibilities to learn new and more complex things. But for him, it is the quality, and not the quantity of experiences that count the most in the quality of our learning. Also for him, not all experiences are positive in terms of producing new learning. For example, a child might have such a traumatizing experience with water in her first swimming lesson that she might never want to go back to a pool again, which will prevent her from learning to swim.

While Dewey did believe that the curriculum should be based on issues that children are interested in and that are relevant to their real lives so that their school experiences can be meaningful to them, he also defended the idea that we as adults—because we are more mature and experienced than children—should carefully guide them so that they can have truly educative experiences. This, according to him, is extremely important so that children can keep going beyond their present demands, needs, and perspectives, and move towards more complex levels of learning and development.

Another author who has addressed the issue of freedom and structure and who is often credited by holistic education scholars as a significant contributor to the theory of holistic education is Maria Montessori. From Montessori's (1964, 1967, 1966) perspective, one of the main roles of the teacher is to guide children to independence. However, from her viewpoint, the individual can only attain complete freedom to the extent he/she has enough knowledge and abilities to make intelligent choices. Montessori developed a method that grants children freedom to make choices and decisions about their own learning. But in her method, the child can only make choices within a highly structured environment carefully planned by the teacher and shaped by self-teaching materials. In Montessori education, the role of the teacher is to carefully observe the child and interfere with her activities only when necessary (e.g., to show the correct use of a specific material, to teach a new lesson, or to scaffold the child's learning). It is important to note here that, while the Montessori method grants children freedom of movement (they can freely move around and choose the activities they want to engage in), children's choices in a Montessori classroom are limited to learning experiences previously arranged by the teacher and to prescribed ways of handling the instructional materials available in the classroom.

Another author in the holistic education literature with a clear position on the issue of freedom and structure is Rudolf Steiner (e.g., 1920/1995, 1995, 1998). According to him, children are born with inherent capacities that require nourishment and discipline so that they can develop to their fullest capacity. Ron Miller (1995) summarizes the essence of Waldorf[3] education as it pertains to the issue of freedom and structure:

> The Waldorf educator sees unbounded freedom for young children as premature, as violating their innate developmental needs. The Waldorf approach is grounded in the understanding that the growing child possesses an unfolding spiritual identity, which can be nourished with loving yet firm guidance. According to this developmental

> view, human capacities are inherent, but they require cultivation and discipline in order to be expressed to their fullest measures. (p. 8)

As I will extensively talk about Steiner and the Waldorf perspective in the second section of the book, I will not extend myself in this subject here.

A different perspective of freedom is well illustrated by the work of the twentieth century Indian philosopher/sage, Krishnamurti. Freedom for him is related to the idea of absence of conditionings and fears. From his perspective—actually from the perspective of most Eastern philosophical traditions, as well noted by Yoshiharu Nakagawa (2000)—we necessarily need to get rid of the hindrances of the mind (e.g., pre-established and fixed theories, patterns of thought, and fears) and from all sorts of attachments we might have, if we are to have a clear view of reality and if we are to be truly free. Freedom from this perspective comes from self-knowledge and self-awareness. It is not necessarily related to absence of external constraints, political power, or to the ability to do what one wants to do. Krishnamurti (1981) believed that we as educators should not impose our agendas, values, goals, and knowledge on students in order not to further condition them with our own preconceptions and worldview. He was pretty much against the use of authority in education. According to him, while authority might generate a sense of safety, it might also prevent students from having to face their conditionings, limitations, fears, pains, and attachments. From his perspective, teachers should only attempt to guide students to the extent that they are sure they do not have any desire to control or direct them toward particular goals. Since he believed so, he did not prescribe any specific educational practices.

The holistic education movement as a whole does not offer as clear a view on the issue of freedom and structure as do many of the authors mentioned above, except for the idea that we need to find a balance between freedom and structure according to contextual factors. Ron Miller (1995) explains:

> From a holistic perspective, freedom is always situated in a particular context involving the particular temperaments and developmental levels of the persons involved, the particular educational task at hand,

> and the particular social, economic, and cultural realities being experienced. A holistic educational environment is not fixed around a single notion of "freedom" but remains flexible and responsive to its changing situational context. (p. 8)

According to Miller, a holistic perspective in education calls for a certain degree of freedom, so that children can pursue their interests and explore their unique potentials, yet it also calls for a certain degree of structure for three main reasons: First, as Dewey (1963), Montessori (1966, 1964, 1967), and Steiner (1995, 1998) advocated, children need a certain degree of structure and discipline to support them in developing and learning to their best abilities. Second, considering that one of the main goals of holistic education is to promote harmonious and healthy relationships between us and other people, and between us and the world around us, we need to teach ourselves and our students to establish limits to our personal demands. Third, but not least important, true freedom requires discipline so that we can find our true self and not get entrapped by the values and demands the dominant culture imposes on us.

The lack of a clear standpoint on the issue of freedom and structure in the holistic education literature is not necessarily problematic since the holistic education movement does not attempt to prescribe specific educational practices. However, considering that in reality many people are trying to implement holistic education ideas in practice, forums of discussion on the complexities involved in this issue are necessary. This brief review on freedom and structure in the context of holistic education was meant only to introduce the subject, which will be thoroughly discussed later in the book. But note that from now on, my approach to this matter will be based on the perspective of the research participants on this issue rather than on the viewpoint of the different authors.

Notes

1. The view of holistic education presented here comes mainly from a Western perspective. In the Eastern world, as Nakagawa (2000) well writes about, holistic education is centered around the idea of helping people realize different levels of reality and achieve enlightenment. For the purposes of this work, I will adopt only the Western perspective.
2. This journal is now called *Encounter: Education for Meaning and Social Justice* and has been since the Spring of 1988 the main forum of holistic education ideas.
3. Waldorf education is an educational approach that closely subscribes to Steiner's educational philosophy and practice.

The
School Around Us

Chapter 1

The School Around Us

A BRIEF HISTORY OF THE SCHOOL

Inspired by the social and educational criticism of the 1960s, many groups of parents gathered around the country to discuss the problems related to the education of their children, hoping to come up with a better educational system for them. This was the case of a group of parents residing in the Kennebunk area of Maine. They wanted their children to be able to be themselves and to follow their individual interests throughout their school experiences. They wanted them to be able to keep their curiosity and interest in learning. They also wanted to ban the tracking system, which was undermining the self-esteem and self-confidence of many of their children (Berman, 1994a).

At first this group of parents tried working with local authorities to change the school system. But as this didn't seem to be leading them anywhere, they decided to start a school of their own. Two of the parents, Stacy and Marilyn Wentworth, donated a generous piece of land adjacent to their farm for the school. The group of parents then got together to build the school, which they did, step by step, with their own hands. Since the foundation of the school in 1970, the Wentworths' farm—which is actually an organic farm—is treated as a natural extension of it. The teachers and children often go there for specific learning activities and also to observe animals, to hike, bike, swim, canoe, do cross-country skiing, and to ice-skate on the pond. In fact, not only the farm, but everything around the school is treated as an extension of it, which is one of the reasons why the school was named "The School Around Us" (I will explain this further later).

It is interesting to note that the group of parents who started the SAU wanted to create more than a school for their children. Like many other groups of parents who were starting free schools all around the country, they wanted education for themselves and looked forward to effecting radical changes in their lives. When they got together in meetings, they discussed not only details related to the school, but also talked about issues such as how to grow organic food, start cooperatives, use energy efficiently, and reuse and recycle materials.

The SAU started in September of 1970, with 30 students, ages five to thirteen. The enrollment was limited to 30 students so that the school would not be too big and the teacher-student ratio would be high. The school looks more like a house than a school as most of us know schools. It has three levels directly interconnected, and in each of these levels, there is a large room used for classes and for school meetings. Each of these rooms is surrounded by large windows, which maximize the use of natural light, and give anyone inside these rooms a generous view of the beautiful woodland outside the school. In the intersection between the three main rooms, there is a kitchen where people gather to cook, eat their meals, and talk. Adjacent to each of the main rooms are little rooms which are used for storing materials or as special places for computing or developing special projects. Besides, each of the rooms has little cozy nooks where children can hang out by themselves or with a couple of friends for some privacy. When the school year started, the school was not yet completely built (it did not have doors or running water), and helping to finish building it was considered a part of the curriculum.

How the School is Run and Organized

The Cooperative

The SAU is a democratic cooperative run by parents, teachers, students, and friends of the school. The members of this cooperative participate in school meetings and serve on one or more committees that take care of all business of the school (e.g., fund-raising, maintenance of the school facilities, recycling, and organization of open houses). During the academic year I did my research (1997-1998), the parents met every Thursday evening for about two hours to discuss administrative and sometimes, pedagogical issues. The teachers were welcome at these meetings, but didn't always participate in them. All decisions at the parents' meetings were made by consensus, since the school community believes that consensual decisions lead everyone to become attentive to, and responsible for, everybody else's needs.

The students also shared some responsibility in running the school. Even though they did not participate in the parents' meetings (except in special circumstances), they actively participated in daily and weekly school meetings, where along with the teachers, they discussed issues of concern to the whole school community and made decisions about upcoming events and learning activities.

Some friends of the school—ex-teachers, ex-interns and alumni and alumnae—sometimes also volunteered to help with school projects (e.g., fundraisers, cleaning up the school, organizing mailing lists) and/or to participate on the school board. The school board did not make administrative decisions. It met a few times a year (three or four times) to discuss issues related to the school's philosophy of education and how it was actually being implemented. It also organized annual school meetings that included all the people who were—or who one day were—connected with the school.

The School Budget

The SAU is a non-profit cooperative. The total budget of the school depends on the payment of tuition and fund-raisers. The tuition is low when compared with that of other private schools. During the year I observed the school, while most parents would pay somewhere between $6,000 to $10,000 to have their children in private schools, the SAU parents on average paid less than $3,000. Tuition was pledged on a form and only the treasurer knew what individual people had pledged.

In order to raise funds, the school organized three crafts fairs a year. These fairs were famous in that region of New England for the quality and diversity of hand-made arts and crafts sold. They generated more than half of the total budget of the school through fees collected from the artisans who were selected to have a booth space to sell their products.

Aside from the crafts fair, the SAU community eventually organized some special events such as yard sales, suppers or plays by the children to raise money for specific projects planned by the teachers and students (e.g., field trips to distant and/or expensive places). When this was the case, the students themselves managed the money until they could use it in their activities.

The School Schedule

The structure of the schedule of the SAU has varied a lot through the years depending on the demands and needs of the parents, teachers and students. During the initial years of the school, there was hardly any schedule at all. It was only in 1978 that the parents decided there was a need to have some kind of schedule, even if a loose one, to give some structure to the day of the children. During the year I observed the school, the days started with daily morning meetings where the teachers and the children talked about issues of concern (e.g., the news someone heard about, the trip that one of the children took to his/her grandmother's house, the death of someone's dog, the new hair style of one of the students, what they would be doing during the day). These meetings usually ran from 9:00 a.m. to 9:20 a.m. or a little later. When they finished by 9:20 a.m., the remaining time was used for "all-school games" which were put together by the students themselves.

After the morning meetings or all-school games, on Mondays, Tuesdays and Thursdays, there were two blocks of classes (from about 9:30 to 10:30 a.m. and from about 11:00 a.m. to noon) with some time for a snack break between them. One block was dedicated to math and the other to language arts. The students rotated in attending the math and language arts classes along with their peer group (there were two groups: younger students and older students). At lunchtime, they had one hour to eat and play. After 1:00 p.m., they would work in different classes or on theme projects individually chosen by them. On Thursdays, before their normal afternoon activities, from about 1:00 p.m. to 1:45 p.m., they had "all-school meetings," where they discussed issues of common interest. On Wednesdays, they started the day with a school meeting and then had arts for the rest of the morning. During the afternoon, they had community service. On Fridays, after the morning meeting, they had sports and free activities for the rest of the day. The SAU schedule was meant to offer some structure to the day of children. It was not meant to be rigidly followed, and therefore, it was often changed to accommodate the emerging needs of the school community.

The Educational Philosophy

The name of the school—School Around Us—stands for the idea that the world around the school is also a school. Therefore, for the SAU community, teaching and learning processes should treasure not only the learning experiences that can happen within the school building, but should also include and treasure the learning possibilities inherent in the school around the school, the world around us. The holistic philosophy of the SAU is translated into practice, according to the school handbook (Board of Trustees of the School Around Us, 1997), through four main educational goals. Following are some passages from the school handbook where the school community explains how they promote each of these goals:

Goal 1: Learn how to learn, for life by…

- respecting choice as central to learning;
- exploring and practicing basic skills;
- applying basic skills to real life, hands-on, learning experiences;
- fostering the pursuit of genuine interests and passions within the classroom and extending into the community;
- making use of community resources, including mentors, learning sites, apprenticeships and genuine work;
- encouraging peer learning and appreciating of difference;
- encouraging the pursuit of beautiful, challenging, useful work;
- appreciating the value of mistakes. (p. 2)

Goal 2: Discover your whole self, by…

- engaging in activities of the heart, mind and body;
- paying attention to learning styles and various intelligences;
- practicing reflection and introspection;
- playing and laughing, dancing and singing;
- encouraging each child to discover his/her own artist within;

- developing, and discovering, individual skills and interests. (p. 3)

Goal 3: Live in harmony with your natural environment, by…

- practicing reduce, reuse, recycle every day;
- thinking globally, act locally in terms of responsible, sustainable living;
- exploring the ecosystem of our 'back yard' and our role in that system;
- exploring the interrelationship of all things;
- actively participating in public issues that effect the natural environment. (p. 3)

Goal 4: Practice responsible, knowledgeable citizenship by…

- encouraging the development of personal values, ethics and a sense of justice;
- practicing respect, appreciation, compassion, and acceptance of diversity;
- using community meetings and consensus to make decisions about learning, justice, governance, and relationships;
- accepting responsibility for one's own actions as they affect other individuals and the immediate and global community;
- engaging actively in multi-cultural experiences;
- participating in and contributing to the social, political, environmental and economic life of the various communities that touch the lives of the learners;
- considering all sides of issues;
- exploring the benefits of diversity in the human and physical world. (p. 4)

SAU Students and Teachers

In 1998, when I collected the data for my research, the school was going through a period of transition and was actually putting together many open houses in an attempt to enroll new students. At that time the school had only ten registered students (ages 6 to 13). Six of them—Abiel (11 years old), Casper (6 years old), Clayton (8 years old), Claus (12 years old), Elise (10 years old), and Shandy (10 years old)—had never gone to another school. Their parents fully believed in the SAU educational philosophy and were quite committed to it. Two students—Nathaniel (13 years old) and Karl (12 years old) enrolled at the SAU in the academic year 1997-1998 because they were having problems in the public school system and wanted to try some kind of alternative education. Two other students—James (11 years old) and Allyson (10 years old) enrolled in the school because the SAU seemed to be the best option available. James had been studying in a similar alternative school in Santa Fe before he moved to Maine, and his mother wanted to keep him in an alternative and small setting. In the case of Allyson, her parents felt attracted to the SAU philosophy, and also were homeschooling her and wanted her to attend school only part time, which was something that the SAU did not have any objections against.

When I collected the data for the research, the school had three teachers: One full-time, Greg Rogovin, and two part-time, Amy Wentworth and Kendall Temper.

Amy Wentworth. Amy started attending the SAU even before it started, since her parents, Stacy and Marylyn Wentworth, were two of the main founders of the school and were quite involved with it. She studied at the SAU from early childhood to the end of elementary education. Her four siblings also attended the SAU for most of their elementary years, and her mother and three of her siblings actually taught there either as regular or substitute teachers.

When Amy finished her formal teacher training (she completed her Master's degree in teaching), she decided to work in the Vermont public school system as a first and second grade teacher for five years. She wanted to share some of her alternative background with public school teachers and students. When she got pregnant with her first child, however, she decided to apply for a teaching position at the SAU. She could not afford stop working when her child was born, and she knew that this school would probably allow her to bring her baby to work with her. At the end of the first semester, she took a leave of six weeks and had her child, Alexander. After that, she returned to the school with him, who from then on, became an integral part of the school community. After Alexander was born, Amy, who used to work full time, decided to go to school only part-time. She and the parents group invited Kendall, who had substituted for her during her maternity leave, to permanently take her place in the language arts program three mornings a week.

Kendall Temper. Kendall's first experience at the SAU was as an intern during the first year of the school, in 1970. She taught again at the school for one year in the 80's, and then again for two years during the 1995-96 and 1996-97 academic years. During the year I observed the school—1997-98—she was hired in January to substitute for Amy during her six-week maternity leave. However, as I just explained, she ended up staying at school part time as the official language arts teacher until the end of the school year.

Kendall has been attracted to alternative approaches to education since her youth and felt at home in the SAU. After all, some of the authors that most influenced her educational philosophy—A. S. Neill, Jonathan Kozol and John Holt—were some of the same authors who inspired the philosophy of the SAU from its very beginning. While Kendall had graduated as a teacher, she had little teaching experience in the sense of being full-time lead teacher. Most of her work as an educator had been as a teacher of her own children (she homeschooled two of them) and as a temporary substitute teacher in different schools. Her main means of making a living has been various craft work.

Greg Rogovin. Greg learned about the SAU through Claudia Berman and her book *The School Around Us: 25 Years.* As soon as he learned about the school, he felt attracted to it and applied for a teaching position there. He started at the SAU in September of 1997 as an assistant teacher, in spite of the fact that he already had experience as a lead teacher in the third, fourth, fifth and seventh grades in Maine public schools. He put himself in this situation, since there were no full-time positions open and he always wanted to teach in a setting that was committed to holistic education.

As soon as Greg started teaching, both he and Amy realized that he was too dynamic to be just an assistant. Therefore, they started sharing responsibilities. They divided the group of children into two groups: the younger group and the older group. They then split the morning into two halves: while Greg would teach one group for the first part of the morning, Amy (or Kendall, after Amy decided to teach only part-time) would teach the other group. In the second half of the morning, they would change groups. Greg focused on teaching mathematics, and Amy (or Kendall when Amy took her maternity leave and started working only part time) focused on language arts. During the afternoon, Amy and Greg worked in different projects with individual students, small group of students, or with all the students.

I explore the school's educational philosophy and practice more fully in the following chapters.

Chapter 2

The Pedagogy of Connections at the SAU

In this chapter, I will talk about what I came to call the "pedagogy of connections" at the SAU. I will talk about: (1) how the teachers connected the curriculum with the real world as a school from which they and the students could learn; (2) how they connected the curriculum with the students' emerging interests; (3) how they connected with students on a personal level; (4) how they connected with students' strengths as a means to foster their self-confidence; and (5) how they promoted a sense of connectedness between the students and the natural environment.

Connecting with the Real World: The School Around the School

For the SAU community, school and life are not, and should not be, looked at as two separate things. Amy expressed this thought well during two of my interviews with her, as shown in the following quotes:

> We should have school be a part of life and not something that is separate, like creating a mock grocery store and setting up a pretend envi-

> ronment in the classroom ... don't do that ... let's have the environment, let's be in the environment ... go and be in nature instead of trying to set up mock nature in your classroom ... have the experience ... be real to life and not something that is recreated so that students really feel it in a real way ... and to not have the kids have this image of school as something separate from what their real life is.
>
> ... and then trying to have the academic pieces also be connected to real life things, to be real things and not created, you know, fabricated in the classroom.... Why pretend that we are having the real world? Why not give them the real world?

Since the SAU teachers saw the world as a school from which they all could learn, their teaching included and treasured not only the learning opportunities that happened within the school building, but also included and treasured learning opportunities that could be generated as a result of the school connections with different places, people, and situations not officially related to the school. They organized and went on many field trips with the school community. The school's philosophy of education and organizational structure facilitated these trips for them, since: (1) the school had a van that could be used at any time by the teachers for field trips; (2) the teachers had prior approval of the parents to go out for daylong field trips whenever they and the students decided to; and (3) the student-teacher ratio was low, which meant that the teachers didn't have to be too concerned about safety issues when outside the school. Besides trips to places such as different towns, public organizations, shelters and parks, they also organized many little trips to the woods and to the beach. Kendall actually would take the older group either for walks in the woods or on the beach almost every class. She thought this was positive since according to her, it inspired children's writing, fostered in them a sense of appreciation and connection with the natural environment, and allowed them to "let some of their retained physical energy out."

The members of the SAU community treasure the world around the school as well as different people and emerging opportunities as schools from which they can learn. This is so true that even my presence at the school was seen by them as a learning possibility. When I told the school community about my intention to develop a research project on the school, they fully welcomed it, since they felt that, among other things, they would learn about themselves from my research and, by watching me conducting my study, they and the children would learn some things about research methods.

Another great example of the extent to which the members of the SAU community value people and emerging opportunities as schools from which they can learn can be seen in the fact that the parent group not only hired a teacher whom they knew was pregnant, but they supported the idea of bringing the baby to school with her. Instead of looking at Amy's pregnancy and the presence of a newborn in the school as problems—as issues that would disturb the natural flow of school activities—they saw her pregnancy and baby as an invaluable educational opportunity for everyone in the school community. They believed that this experience would teach children a little bit about pregnancy, and about how to take care of and how to nurture a baby's growth and development. It would also teach them that the presence of babies, and all the care and nurturing that babies require, are a natural part of life which can enrich everyone's existence.

Note that the idea of having a baby around in the school seemed to be natural to the whole school community. The students didn't seem to mind when Amy had to change his diaper or nurse him while she was teaching. They also didn't seem to feel too disturbed when Xander cried. Both Amy and the students would do their best to calm him down, and then would try to go back to work. As Xander grew, his presence and his noises became a natural part of the school scenario. Even I became so used to his noises and demands, that from time to time I would have to make a conscious effort to observe how his presence affected the learning environment.

While Amy's, Kendall's and Greg's pedagogy of connecting with the real world, with different people, and with emerging learning opportunities enriched the SAU curriculum and promoted meaningful and purposeful learning, it also generated some problems, which I would like to discuss here.

As teachers and students attempted to connect with the real world by actually going to different places, the time spent in transportation and/or exploration of new environments was sometimes so long, that there was little time left for focused learning. For example, once a week, for about one and a half months, Kendall would take the older students to the beach for their language arts class. She and the students would spend about 15 minutes to get to the beach (this included the time they took to get into the van, drive to the beach, and park the van). The same amount of time was spent getting back to the school. Also, when she and the students arrived at the beach, they would usually spend at least 15 minutes exploring and playing around before they started doing some academic work. In summary, in a class of about one hour, about half an hour was spent on transportation and at least 15 minutes was spent in the exploration of the new environment (or in play). It is true that, as they were going to the beach and as they explored the shore, there was a great deal of socialization, fun, and spontaneous learning (e.g., the children learned about traffic laws, different kinds of vegetation, tides, ocean life, and rock formation). However, during this time, they did little focused reading and writing.

In some circumstances, as the SAU teachers and students connected with different people, they "disconnected from a goal in mind." For example, while Xander's presence represented a lot of learning for everyone in the SAU community, on a few occasions, he became so demanding that it was hard for everyone to concentrate on the schoolwork. However, I do believe that Xander's presence in the school during his first months of life brought more benefits than drawbacks for everyone in the school, especially because he was relatively quiet (he slept a lot) and could be easily supervised. However, I wonder if this would still be true if he were a little bit older, and therefore more talkative and mobile.

Connecting with different people led the SAU teachers and students to disconnect from their ongoing projects in other situations, such as when the school had an open house or received visitors. During the time period I was at the school, there were two open houses and one visit by a group of elderly women, the Golden Girls, who spent an afternoon at the school. Both during the open houses and during the visit of the Golden Girls, the students put aside their ongoing projects and developed special activities. This by itself did not represent a problem, since the students did learn from the special activities and from their relationship with the visitors. However, this started to become a problem when "disconnections" from goals in mind and from ongoing activities started to become frequent.

Based on these observations, it seems that, as children connect with numerous and always emerging learning opportunities from the real world, they might have difficulty focusing and getting deeply into specific activities. And while moving from one activity to another, they might get disconnected from the deeper learning possibilities of each of these activities. Therefore, it appears that sometimes it is necessary to limit the quantity of connections in favor of the quality of connections to be made.

Connecting the Curriculum with Children's Emerging Interests

The SAU teachers tried to promote the knowledge and skills they wanted students to develop through topics and activities in which students had a current interest. For instance, when Greg, the math teacher, decided to have students work on computation skills, instead of having them do the math in their notebooks, he created activities that dealt with issues that the children cared about. For example, as he knew the older students were interested in money and investments, he developed an activity in which children received fictitious money to invest in the stock market. As the children "invested" their money and followed the news on the market on a daily basis, they had to do a lot of calculations about their profits and losses and make decisions about how they could best invest the money they had left.

Amy, Greg and Kendall found many difficulties as they tried to incorporate things in which children were interested into the curriculum. I will address some of these difficulties by elaborating on the example of the "theme project," which came about as a result of students' demands. A few classes before they finished two prior projects they had been working on during the afternoon classes, they asked Amy and Greg whether, for their next unit, they could individually choose a project in which they were interested and on which they could work either individually or in small groups according to their preferences. The teachers agreed with the students' requests and set only one major requirement: Their projects had to have an "academic piece" that should be presented in written form by the end of the semester.

The children chose a variety of projects to work on: Allyson (10 years old) and Elise (10 years old) conducted research on monkeys (each of the girls studied a different kind of monkey). James (11 years old) did some research on the life and habits of ferrets. Claus (12 years old) and Nathaniel (13 years old) built a golf course in the schoolyard. Clayton (8 years old) did some research on how to play golf. Abiel (11 years old) and Shandy (10 years old) built an obstacle course around the school building. Karl (12 years old) developed a rope course in the schoolyard. Casper (6 years old) made a little book on different kinds of animals. While the students made decisions about how to develop their projects, and while teachers tried to respect their choices and tried to help the students according to their needs, both teachers and students faced many problems. For example,

- James took about two weeks to decide what he wanted to do, which meant that he ended up with a limited time slot to actually work on his project. He is a child who needs some extra time to make decisions. It would have probably been better for him if he had known in advance that he would be working on an individually-chosen project. This way he would have had more time to make his decision on time and would have been ready to start working at the beginning of the learning unit.
- Clayton wanted to build a golf course with Claus and Nathaniel, but those two didn't want Clayton in their group. It was hard for the teachers to mediate this problem considering that, when the students had asked them for permission to do the theme projects, they said that the students could work either alone or in a group of their choice. Having said that, they could not force Claus and Nathaniel to accept Clayton in their group.
- Some of the students, more specifically Abiel, Shandy, Claus and Nathaniel, were more interested in the practical than in the academic aspect of their projects. In other words, they were more

interested in building their obstacle and golf courses and in playing on these courses than in learning new things about these sports. This led them to set low standards for the academic piece of their project. The teachers tried to prevent this problem by having the students fill out a form called "proposal for independent study," where they asked students to write about what they were planning to do in order to learn new things in each of the different academic areas. In spite of the teachers' efforts to have students plan and set high standards for their academic work, some of them gave only easy answers to the questions asked. For example, when Abiel and Shandy got to the part of the form where the teachers asked them what they would do to promote mathematical learning, they wrote that they would time people to see how fast they could complete the obstacle course. They also wrote that they would estimate the costs of the materials they would need for the obstacle course. Considering Abiel's and Shandy's age, they certainly could have asked much more of themselves in terms of their use of mathematical skills. Both Amy and Greg worked hard in trying to raise students' standards. However, the students resisted so much that, at the end of the project, they were happy to have the children at least complete the assignment.

- Because Abiel, Shandy, Claus, and Nathaniel were more interested in the practical than the academic part of their project, they procrastinated starting with the academic work that their project required as much as they could. Even though the teachers spent class after class talking to them about the importance of this kind of work, they kept on finding excuses not to do it. When the teachers started to get harder on them and tried to make concrete suggestions about what they could do for the academic piece of their project, they rebelled and said that the project was theirs and nobody else's.
- Because the students had different interests and were working on different projects, it was hard for the teachers to give individual attention to all of them. Also, because both teachers worked with all the students and students worked in different places in the

> school (outdoors and indoors), the teachers easily lost track of what each student was doing, and which of them had already received attention from them or from the other teacher. As a consequence, some of the students ended up receiving less attention than others.

Based on the experience described above, apparently some of the problems teachers have to deal with while trying to develop the curriculum according to children's emerging individual interests and needs are: (1) As teachers connect with students' immediate interests and needs, they might not have enough time to plan in advance to provide for the students who need extra time to make decisions and get organized. (2) Many children have a tendency to be more interested in the practical than in the academic aspects of their learning experiences. If we do not create some attractive strategies to get them interested in academic activities, they might consistently get away from involving themselves in academic development and keep missing learning opportunities in this area. (3) The interests and needs of some students might conflict with the interests of others. If the teachers don't have good strategies so that these interests are fairly negotiated, the most vocal and aggressive students might always get things their own way. Moreover, if the teachers do not have a system in place to address the needs of all students, they might end up devoting more time and energy to the "loudest" and most problematic children.

Three additional issues also stand out from my observations: First, children lack experience to know exactly what skills and what kind of knowledge they need in order to do well in the world when they leave school. Second, children change their immediate interests a lot. One day they might be interested in doing one thing, and the very next day they might be interested in something else. As a consequence, they might keep moving from one interest to the other without going deeply into any of their projects. Third, what children want to do is often influenced by what they know. The more they experience rich learning opportunities, the more likely they will be to want to invest themselves in increasingly complex learning.

Connecting with Children on a Personal Level

Amy, Kendall, and Greg tried to develop a close and personal relationship with students. For instance, whenever Amy and Kendall started a conversation with them, they would try to be completely present and involved in it as if there were nothing more important for them to do. They would carefully listen to children, promptly answer their questions and demands, and share personal facts of their own lives that related to the subject of the conversation, even when these subjects, from the perspective of most adults, seemed unimportant for a teacher to get involved in. Once, for instance, we were in the van coming back from a field trip to Portland, Maine, and Shandy (10 years old) started telling Amy about a trip to New York he would take in two weeks. As Amy listened to Shandy telling her about his trip, she started asking him questions about what he was going to do in New York, who was going to go with him, what places he would visit, and why he was so excited about the trip. As he answered her questions, she also shared with him that in a few weeks she too was going to travel with her husband and child and that she was really excited about it. They got fully immersed in this conversation until another child demanded Amy's attention. The kind of attention Amy gave to her students brought them close together and very much contributed to their relationship.

Let me tell a little story about Kendall to illustrate the extent to which she too, like Amy, deeply connected with students. On a beautiful spring day, we went on a hike in the woods behind the school and across the train tracks. After we had walked for about 20 minutes, we sat in a spot by the river and Kendall asked the students to write about anything they would like. The students wrote for a few minutes, and, when they were done, they restarted their hike. When I was about to follow them, I realized that I had lost my glasses somewhere in the woods. I asked Kendall if we could go back to the school the same way we had come, so that I could look for my glasses. As the students were a little bit ahead of us walking towards the road, she told them she was going to go back to the school with me and that if they wanted, they could go back to the school some other way. When Kendall and I got back to the school, the students who were already there asked us whether, while we were in the woods, we had heard someone yelling on the road and we said we had not. Abiel then told us that when they were coming back to school, he spit at a truck which passed by on the road. The driver got so angry that he backed up to yell at him. The children got so scared that they all ran away and hid in the woods while the angry man yelled at them.

When the students finished the story Kendall laughed and said: "And Dora had just asked me whether I trusted you and I said I did! And now you guys do that!" She then asked them to think about what they had done and about the consequences of acts like that. But she did not scold them. She knew that the students had already learned a good lesson when the truck driver backed up and yelled at them. Students were not afraid to tell Kendall about things they had done wrong because they saw her as a nonjudgmental friend, and as she was usually so honest with them about her own life that they responded with honesty. Once I heard one of the students asking her whether she had sneaked into concerts and smoked marijuana when she was younger. Instead of ignoring the question or denying the fact in order to keep up appearances, she gave him a simple, honest and direct answer. She said yes, she had done it in special cases. As students saw Kendall as a normal human being who also had her "weaknesses" and as a good friend whom they could trust, they shared with her secrets they wouldn't usually dare to share with other adults.

Like Amy and Kendall, Greg too tried to connect with students on a personal level. However, he was so concerned that some of the students were not doing as well as they could in their academic development, that he would usually devote more time to this area than to his relationship with them. As Greg wanted to have students work harder than they wanted to and were used to, and as he tried to implement some rules to deal with behavioral problems, the students little by little started resisting his person, ideas, and demands (I will talk more about this subject in a later chapter). By the end of May, some of the students were so upset with him, and his relationship with them had become so tense, that he had to significantly lighten his demands for their work and behavior. Usually Greg would have a good and more personal relationship with students when he spent time with them individually, away from other students.

Connecting with Students' Strengths

The three SAU teachers talked about fostering students' strengths and confidence in themselves as learners as one of their educational goals. In the following paragraphs, I will give some examples of how each of them attempted to promote this goal.

Amy, for instance, like the other teachers, faced a challenge dealing with Nathaniel (13 years old). He went through serious behavioral problems: He constantly refused to engage in activities proposed by the teachers. He also disrupted the flow of classes by talking loudly to his peers, by unexpectedly leaving the room, and by making jokes. But, in spite of all of this, Amy was always trying to have a dialogue with him, to bring out his positive side, and to optimize his strengths. In an attempt to get to know him better and to improve their relationship, she invited him to write what she called an "interactive journal." In other words, she suggested that he share his journal with her so that she could also write in it and they could communicate in this additional way.

As Amy read and responded to Nathaniel's journal entries, she never made any negative remarks about him or about his work or behavior. All her comments were either an attempt to get to know him better, to foster a closer relationship with him, to congratulate him on his strengths and/or to challenge him to use some of his strengths—his organizational and leadership skills—to help her do something.

Amy believed in the goodness of everyone and was always trying to capitalize on this goodness. For example, a couple of months before the semester ended, Nathaniel asked her whether she could hire him to work with her during the summer camp she was organizing. Considering that Amy knew about Nathaniel's behavioral problems, she could have easily rejected his offer, but she didn't. Instead, she told him that she thought he was very good with young children and therefore, she would love to have him help her. The only comment she made regarding his behavior was that she didn't want to have to waste her energy on it. She believed so much that, if given a chance, he could do well, that she not only accepted his offer to assist her at the summer camp, but she also invited him to help her plan the camp activities.

Like Amy, Kendall was also dedicated to promoting the students' strengths. As she stated during one of our interviews, two of her main goals as a teacher are to help students "become aware of the things they are good at" and to "bring out the best that is in kids." Kendall was in fact so focused on reinforcing and promoting students' strengths in order to foster their confidence in themselves as writers that she would rarely comment on the weaknesses of their work. She wanted them to feel successful:

> I want every child to think of herself as a writer, as a reader, as a person who is capable of doing all these things. That is one of the main things, you know. It's like, "yes, I can write!" So whenever they need to write something, they just know they can do it because they had the experience and they have been successful, and the same with reading.

Like Amy and Kendall, Greg too focused more on students' strengths than on their weaknesses. For example, one day, he had each of the students present the findings of his/her interviewing project (for this project, each child was supposed to interview someone on how he/she used mathematics in his/her work). After the presentations, one of the students raised his hand and asked Greg to comment on how each student had done in his/her presentation. Agreeing to the student's request, Greg started talking about the highlights and positive aspects of each presentation. It was only when some of the students asked him to talk about the weaknesses of their presentations that he decided to address them. But he did so hesitantly and avoided naming names. Whenever he pointed out problems, he spoke as if he were talking with everybody, and not with any individual student. By doing it this way, he didn't put any of the students on the spot, nor did he reinforce their particular weaknesses by repeatedly pointing them out.

As in the situations mentioned above, Amy, Kendall and Greg also focused on students' strengths as they wrote their end-of-year evaluation letters. In these letters, whenever they wanted to bring up something that was not exactly a strength of the child, instead of stating it as a weakness, they would use words such as so and so is "developing sense of..."

or "is still developing in...." They made very few comments on students' weaknesses, and whenever they made such kinds of comments, they used some kind of strategy to lighten their criticism. One of the strategies they used was that, either right before or right after their comment, they would say something positive about the child. Another strategy they used was to try to explain the reasons they had that specific problem, as if trying to protect them from feeling guilty. For example, in her evaluation letter to Casper (6 years old), Amy pointed out that he seemed not to like to participate in "organized theater" and didn't seem to be "fond of singing." But right after she made these comments, she explained that this was the case probably because his peers were "twice his age," which made singing and participating in organized theater "too intimidating" for him.

Before closing this section, I would like to make two remarks, one positive and one negative, of the "pedagogy of connecting with students' strengths." The positive side of this pedagogy is that, as my interviews with students and ex-students of the school reveal, the SAU students tend to have a good sense of what they are good at doing, and usually learn to use their strengths to compensate for some of their weaknesses. The negative side of this pedagogy is that it doesn't necessarily lead students to face and to work on their weaknesses. This might counter the best interests of some of the children for whom the school might be the only supportive environment where they could safely face and work on their weaker areas.

Connecting with the Natural Environment

For the SAU, and for its teachers, the natural world is an extension of the school—a school around the school—which everyone can learn in, learn with, and learn about. They believe that by spending time in the natural world, and by observing and exploring it, one can learn a lot about plant and animal development, the different seasons of the year, soil composition and erosion, rocks and geological formations, the properties of air, water and fire, etc. For them, the natural world can be a perfect classroom even when learning about nature is not the main goal.

But learning in, learning with and learning about the natural world is not enough for the members of the SAU community. For them, it is extremely important that everyone learns to live in harmony with nature. The school has a policy that everyone should reuse and recycle as much as possible. During the time period I observed the school, they separated the trash in special bins for white paper, colored paper, cans, plastic, glass, and organic trash. Whatever could be recycled and reused at the school, was recycled and reused at the school. What could not was taken to proper recycling centers. The teachers and some of the parents took the issue of reusing and recycling seriously. Once, for instance, Greg had some of the children clean out the refrigerator as a "consequence" for their misbehaving. Because everyone, including the children, stored food in the refrigerator, there were many containers full of food that had been forgotten there for a long time. Greg had the children go through each of these containers to sort out what was still good and what was spoiled. He then had them recycle the spoiled food in the compost bin, clean the containers, and recycle them in spite of the terrible smell and in spite of the children's complaints about having to do such tasks. As Greg made sure they reused what was still good and recycled the food and the containers in spite of the smell and complaints, he sent everyone the message that we should reuse and recycle even if, in order to do so, we need to make a considerable effort to go beyond our comfort level.

The SAU teachers and some of the parents tried to have students develop respect for all living things. One funny story illustrates this point. Once I went into the bathroom and when I was about to use the toilet – I actually had my pants down—I saw a snake by the sink, about four feet from me. Quite scared, I pulled my pants up and in just a few seconds, I was back in Amy's classroom telling everybody about the snake. Everyone ran to the bathroom and one of the students got a broomstick to move the snake so that they could check what kind of snake it was. When Amy saw the broomstick she immediately advised: "Please be careful not to hurt the snake." It is interesting to note that she didn't say "be careful with the snake." She said instead "be careful not to hurt the snake." Probably, just by glimpsing it, Amy knew the snake was not poisonous, and this is why her priority was not the children but the snake. The snake ended up slithering away through a crack in the bathroom. No one, neither the children nor the snake, got hurt.

The SAU teachers and students spent a lot of time outdoors. They would sled, ski, climb trees, build tree houses, and swim in the pond or at the beach, according to the seasons of the year. While I was doing my fieldwork, I never saw any of the teachers preventing children from going outside during snack or lunchtime. I also never saw them admonishing the children for getting wet or dirty because they were outside playing in snow, rain or mud. Instead, they celebrated with the children the different seasons of the year, and often also got wet and dirty with them.

Amy's, Kendall's, and Greg's pedagogy of connections nurtured students' self-esteem and self-confidence. It also fostered children's sense of connectedness and interdependence with the natural environment, and promoted meaningful and purposeful learning. But while this pedagogy was positive in these ways, it led to some problems that need to be considered further. The students had difficulties getting deeply into learning experiences. Also, they would constantly miss important opportunities for academic development and would avoid facing and working on some of their weaknesses.

As we consider implementing a pedagogy of connections such as Amy's, Kendall's and Greg's, we need to deal with some challenges. One of the most important, I believe, is finding a balance between (1) being open to emerging learning experiences and promoting long-term goals; (2) attending to children's immediate interests and addressing their long-term needs; (3) promoting concrete hands-on activities and moving to more abstract levels of experience for focused academic work; and (4) nurturing students' strengths and making honest criticisms (without hurting or blaming children for their weaknesses) and giving students support to improve on their weak areas. More research is needed on the complexities involved in these tensions so that we can come to more informed decisions on how to deal with them.

Chapter 3

The Social Curriculum of the SAU

While the SAU curriculum was aimed at promoting the development of the whole child and most activities were meant to encompass all the different aspects of children, students' social development was an essential part of most of what was done at the school. Many activities were in fact designed into the schedule in order to take care of this goal. Among them, there were the "morning meetings," "all-school meetings," "mediated meetings," "group games," "community service," and "clean up" projects.

The morning meetings happened on a daily basis, first thing in the morning for about 20 to 30 minutes. During these meetings, teachers and students shared issues of personal interest and discussed as a group some of the activities that were going to take place during the day. The "all-school meetings" happened every Thursday from about 1:00 p.m. to about 1:45 p.m. During this time teachers and students discussed subjects that concerned the whole school community such as a new project they would like to work on, a new discipline system, the problems they were having with each other, or the upcoming overnight field trip. The students took turns running these meetings and decisions were made either by consensus or majority vote. Only in a few situations when things got too chaotic and the students seemed to be unable to manage the course of the discussion did the teachers moderate the meetings so that everyone could come to some kind of agreement. During both the morning and all-school meetings the students shared and processed important things that were going on in their lives. They practiced talking in public, listening to and supporting each other, making decisions about their experiences, and devising strategies to take action on different kinds of issues.

Besides the regular school meetings, the school also had "mediated meetings." These meetings happened in special circumstances in which parents, teachers, and/or students could not solve problems they had with each other. Throughout these meetings, each of the parties involved had a chance to talk to and be listened to by the other with the help of a mediator. At the end, the mediator helped those involved in the conflict to devise strategies to better deal with the problem under discussion. The mediated meetings not only helped members of the SAU community to deal with problems they were having with each other, but also modeled to them healthy and powerful ways to deal with and solve their conflicts.

The "group games" happened after the morning meetings, for about 10 or 15 minutes, when these meetings ended on schedule. The students took turns leading games that were usually played outdoors and involved physical activity. Some examples of these games were tag, captain tag, blindfold tag, and shooting hoops. The children loved this time of the day when they socialized, played, let some of their physical energy out, and had fun together.

The community service activities were developed on a weekly basis, on Wednesday afternoons. During the time I observed the school, among a few other small projects, the students completed the development of a bridge connecting the Neverdun Farm (the Wenthworths' farm) to the SAU property. They also put up a tetherball pole and a swing set in the school backyard. These projects were meant to improve the school facilities for the SAU community. During other years, according to Berman (1994a) and to the school handbook (Board of Trustees of the School Around us, 1997), besides developing community service projects within the school, they also developed projects outside the school. For instance, they worked on and sold art pieces to raise money for a lodging house for HIV positive people. They visited and helped at old folks' homes and shelters for the homeless. They helped pass referenda that protected the environment (e.g., they helped pass a bill that required that bottles and cans be returnable in the state of Maine, and a bill that prohibited public land from being put up for auction in Kennebunk).

As the SAU students made decisions about how to run the school, lead school meetings and activities, and engage in community service, they developed an increased sense of social awareness and responsibility and practiced participating in democratic processes. Besides, they learned to share and defend their ideas, to consider each other's viewpoints, and to respond to each other's needs.

In addition to the activities that were a regular part of the school schedule, there were aspects of the teachers' pedagogy that also fostered the social skills of the students. The most important, I believe, was that the teachers promoted a lot of collaborative work and consciously avoided promoting any kind of competition between students. Even when the students were not formally working in groups, they encouraged them to share resources and to help each other.

The teachers consciously avoided stimulating competition among students and when for some reason competition was inevitable, they tried to minimize its consequences. There was one occasion on which two of the students decided to participate in an essay competition on patriotism promoted by the state of Maine. The winner would have his essay forwarded to the next level of competition and win a prize of $50. Kendall supported the idea, since these two students were motivated to devote themselves to all the research and writing this project would entail. She thought this would be a good learning opportunity for them. However, as she realized that both students were working hard, but only one of them would win a prize, she got so concerned that she decided to talk about this issue with the parents during their weekly meeting. She and the parents debated this issue for more than 1½ hours! At the end of the meeting they decided that Kendall could take some money from the school fund and give it to the child who finished in second place in order to recognize his efforts and minimize any possible hurt feeling.

"The school felt like a family. If a student had a problem, it was everybody's problem" (John Bordage & Caleb Clark, 1989). This felt true because of situations such as the one described above, and also because some of the SAU members were extremely dedicated to the well being of the school and of the school community. As I collected data for this research, I observed many kind gestures that revealed the extent to which the SAU teachers and parents were attached to, and cared about, the school and the people there. One time I observed Claudia Berman opening the refrigerator door to put in her son's lunch. When she opened the door, she paused and started looking for something that smelled awful. After a few minutes, she figured that probably many of the things in the refrigerator had gone bad because the day before, the kids had turned the refrigerator off to clean it, and forgotten to turn it on again. Even though Claudia was late for work, she didn't hesitate to inspect the food, throwing away what was spoiled, and recycling what had to be recycled. Claudia's gesture, as well as the gestures of many other parents, modeled to students a gentle and collaborative way of being a part of a community.

The SAU teachers often reminded students about the kinds of responsibilities they had as members of a community. Once, for example, during one of the morning meetings, I observed Amy talking to the students about the importance of the "end-of-year school clean up" so that the school would be ready for the next year. She said that she was disappointed in the poor participation of some of the students in this activity, and explained to them that, since they were a part of the SAU community, they had an obligation to help maintain the school facilities. Below is a little excerpt of the conversation she had with the students on that day:

> Yesterday, I was getting frustrated and a little disappointed with some people.... It was hard for me to feel as if we are part of a school and a community and this is our school.... I just didn't feel that some people care about the school. They were complaining a lot like "I'm not going to do this or I'm not going to do that." It isn't always fun to clean ... and care for things but it's a necessary piece. What makes the school unique is that it belongs to all of us. And part of owning something is taking care of it ... we all own the school together: the board, the teachers, the students and whoever is at the school at the moment. Part of being a member of the school, part of your responsibility is to care for

> the school. The parents all have signed up to do some cleaning and care for the school. The teachers will do it and you too. (Audio-taped, 6/1/98)

Besides participating in special activities such as the "end-of-year school clean-up," the students participated in the cleaning up of the school every day before going home. They took turns at different jobs. For instance, during the same week, two children would clean up the kitchen and the bathroom, two children would clean up the library, two children would clean up the upstairs room, and two or three children would clean up the downstairs room. From time to time, they would rotate these jobs.

Field trips were also important means to help the school members build a sense of community and learn what it means to be part of a community. See, for instance, the passage below where Amy emphasized this point during an "all-school meeting," when she, Greg, and the students were making plans for their upcoming camping trip to Acadia National Park:

> Claus (12 years old) - Can I bring my own food?
>
> Amy - No, if you have a problem with what we are serving, you need to talk with the Acadia group …
>
> Claus - Why? What if I want to eat something different?
>
> Amy - You get it when you are at home.
>
> Claus - Why can't I bring my own food and cook it by myself?
>
> Amy - Because this is an event that we are all doing together. Otherwise everyone brings his own food and … cooks their own food in their own little place.
>
> Claus - They should, if they want to they can!
>
> Abiel (11 years old) - I want to know exactly why! [He has a loud and angry tone of voice]
>
> Amy - My reasoning is that I've been on other camping trips, and a certain amount of spirit is built as people are working together to make the meals and clean together. This is our end-of-year trip, and I wouldn't feel comfortable having Claus over in his own little area. He might as well go camp-

> ing on his own as far as I am concerned. So if he wants to come with us, he comes and is part of the whole and cooks and cleans with us. If he wants to go on his own camping trip, he can arrange it with his mom.... (Audio-taped class, 5/20/98)

The fund-raisers, clean-up projects, and field trips were excellent in terms of fostering people's social skills and of building a sense of community in the school. As teachers, students, and parents developed projects and/or traveled to different places together, they learned to work as a group and grew closer as friends.

Before ending this chapter, I would like to talk about two characteristics of the school which had both positive and negative implications for the social development of the SAU students. The first was the small size of the school. The fact that the SAU was small was positive in the sense that, since everyone knew everyone else, it encouraged the development of a sense of community in each member of the school. The downside of the smallness of the school, as Ellen (James's mother) remarked, was that not every student was able to find a "best buddy," which was actually the case for her son (11 years old). Since there were just a few children in the school, and since James's personality did not fit in with the personality of the children who were there, he ended up on his own most of the time. Another problem related to the smallness of the school was that, during the year I collected the data for my research, there were only two girls, Allyson (10 years old) and Elise (10 years old). When one was missing, the other felt a little lonely, and sometimes, also a little out of place.

The second characteristic of the school that had both a positive and negative side was the broad age range of students involved in the same activities. The positive side was that, while the younger students learned from the knowledge and experience of their older peers, the older students learned from teaching, helping, and caring for their younger peers. The negative side was that sometimes the older students overwhelmed the younger students. This actually happened many times during the "all-school meetings," since the older boys were more opinionated, more aggressive, and often had a well-defined agenda they wanted to pursue. Even though the teachers and the mediator of the meetings tried to make sure everyone had an equal right to participate in the school meetings, as the discussions got heated, the younger students and the girls (who were less aggressive) remained quiet most of the time. Casper (6 years old) sometimes would actually leave the room to do something else, since the discussions seemed either "too boring" or did not seem to apply to him.

It is interesting to note that, when I visited the SAU the next academic year, there were more girls at the school and the teachers had separated the students into three groups: kindergarten, primary grades, and middle school. The kindergarten and primary grades would have their "morning meetings" separately from the middle school students. Even though I observed the school for only one day, I noticed that the girls and the younger students were much more vocal than they used to be at the meetings I had observed the year before. Probably one of the main reasons was that they felt encouraged and safer to express themselves with peers of a similar age and/or gender. And, in feeling more encouraged to participate, they were more able to change the agenda of the meetings, making them more interesting for themselves.

Chapter 4

The Big Challenge

Trying to Find the Right Balance Between Freedom and Structure

The theme "freedom and structure" is one of the most important themes that has been running through the history of the SAU since it was founded in 1970. The group of parents who started the school questioned the status quo and called for changes on all levels of society. They did not want an authoritative model of education that would repress the freedom of their children. They wanted them free to fully enjoy their childhood and to question the status quo.

One of the authors who most inspired the parents in creating their own educational philosophy was A. S. Neill in his book *Summerhill: A Radical Approach to Child Rearing* (1960). In this book, Neill defends the viewpoint that "the child is innately wise and realistic," and that "if left to himself without adult suggestion of any kind, he will develop as far as he is capable of developing" (p. 4). Neill also talks about happiness as the main goal of education, and defends the idea that freedom is the best means to happiness, since it is each individual who best knows what happiness means for him/herself and what he/she needs to be happy. As he tells in his book, at his school students had almost complete freedom, being restricted only when strictly necessary to ensure their safety or to prevent them from interfering with other people's freedom. Neill was certain that if we offer children an environment rich in learning opportunities, if we give them freedom to explore their interests, and if we support them in the pursuit of their interests, they will educate themselves and will be more likely to be happy than they would if we directed their learning experiences (Neill, 1960; 1966).

Since the SAU was founded, parents and teachers have been struggling to come to some kind of agreement about how much freedom and how much structure they should provide children throughout their school education. During some years—especially the earlier years of the school, when they had first read Neill—they granted children almost complete freedom to decide what they wanted and didn't want to do. In other years, they tended to be more directive. I interviewed many of the parents whose children were at the school at the time of my research, and they all revealed that they had mixed feelings about this issue. On the one hand, they wanted their children to have a strong voice in their education. On the other hand, they also wanted to make sure their children developed the skills they were likely to need to do well at other schools and in their professional lives in the future.

According to Berman (1994), who explored this issue from a historical perspective, at times SAU parents expected the teachers to be more directive and make the students do more academic work. At other times, they expected the teachers to be looser in their behavior and academic requirements. This inconsistency, she remarks, has generated many discussions that involved parents, teachers and students throughout the years. It has also led many of the parents to remove their children from the school—some because they were not comfortable with the extent of the freedom their children were allowed, and others because they thought their children were subjected to too much structure. The extent of freedom and structure provided to students throughout the school's history, says Berman, has depended—and I would say, still depends—on the combination of personalities and viewpoints of the people who are part of the school community in each specific school year.

Having said a little bit about how the school has been dealing with the issue of freedom and structure, I would like now to focus on Amy's, Kendall's, and Greg's personal views on this issue.

KENDALL

Kendall, along the same lines as Neill (1960, 1966) and other advocates of the free school movement, believes that children can educate themselves and that they will learn when they are ready to learn, when they need to learn, and when they decide they want to learn. Therefore, she doesn't see her role as a teacher to decide what, when, and how students should learn. She sees her role as providing a learning environment and experiences where they can learn according to their own interests, and can feel motivated to always keep learning. She believes in giving children freedom to do the things they like and want to do, to explore their own strengths, and to grow in the directions they would like to grow:

> It [the education of children] has to do with … making sure that there is time and space for each child to be involved in what they love to do and what they are good at doing. So it is like growing a plant. It is providing everything that … the child needs to grow in the best way the child can. And not grow in the direction that you [the teacher] want.… It's that constant search for what is going to work best, what is going to allow this child the most freedom, freedom to grow strong, freedom to grow intellectually, freedom to expand their thinking capabilities, freedom to come to learn what their strengths are so that they just have a chance to be good at what they are good at.

Since Kendall believes that children can educate themselves, she tries to make her expectations meet those of the children and avoids making them do things they don't want to.

> Part of it is about my not having expectations that the kids aren't involved in. If my expectations don't match theirs, then there is not much point in it. It becomes just a struggle, a power struggle, and I am not interested in that at all. So then I have to change my ways.

As the language arts teacher, Kendall wanted the children to love reading and writing. Therefore, she focused more on the process of writing and how pleasurable it could be than on how much and what they actually wrote. Even though she wanted them to read and write quite more than they did in class, and expected them to learn grammar and spelling, while she would encourage them to do so, she would not force them. She trusted that if children learned to like writing, in time, when they were ready, they would learn all the tools they needed to write as these tools were needed.

Even though Kendall didn't ask too much of students, they would often say "no" to her demands and hand in incomplete assignments as they also often did with the other teachers. While she felt disappointed by the students' attitude, she thought giving them the opportunity to make important decisions about their own education was extremely valuable for them.

> I think the main difficulty is that the kids get to say "no" all the time. They have a voice and all, and you sometimes have to spend a lot of time convincing them, rather than just getting on with the lesson, you have to go through explaining why.... But, at the same time, it is just so valuable. It is so important that they are participating in their own education at that level. And even though it makes it harder sometimes, it is worth it. I believe in the process.

During one of my interviews with Kendall, I asked her whether she didn't feel frustrated with children trying to find excuses not to work all the time. She answered that sometimes she did. But she tried not to, because after all, she thinks that children's education is not about her goals, but about children's goals for themselves:

> I only feel frustrated if I have a goal in mind that the child is not meeting. So I try to let go of that. Because it is not about me and my goals. It is about them and their goals. And I don't want to spend my time frustrated over what they are not doing. So I basically try to keep finding ways to have them do what they want to do.

During this same interview, I asked Kendall whether she didn't worry that children were not learning enough for their age level, and not being well prepared to further pursue their education when they left the SAU. She said she didn't worry about it, since she strongly believes that, if the child is exposed to a rich learning environment, she will learn all she needs to learn. And if she doesn't, whenever she needs to learn, she will quickly catch up.

Greg

From his first day at the school, Greg had difficulty dealing with the extent of freedom the SAU students were used to. One of the first incidents in which he had to deal with this difficulty in a more direct way occurred at the beginning of the academic year. He, Amy, the students, and some parents had volunteered to help set up the Common Ground Fair (a fair of organic food producers). At the fair, many of the children did not obey the adults and didn't want to help as promised.

> … We had a six-year-old and older children running around this fairground at night.… I mean, it was a big fairground and I was really concerned, and I was going back and forth to make sure he [Casper] was OK. Steve [Casper's father] was there washing dishes, and I said "Steve, I saw Casper running around out there." He said "yeah, it's OK"… I figured that it was going to be kind of weird, if I was the only one who had major trouble with that.… That was the kind of thing that impressed me the most: the degree of freedom and responsibility that these kids handled or were expected to handle. So I came back and tried to do that here.

Greg said that he wasn't implying that Steve or the other parents were not doing a good job educating their children or that they didn't do a good job taking good care of them at the fair. To the contrary, he had high regard for them because of all he knew about the school. However, he had to come to terms with the reality that the way in which some of the parents dealt with the issue of freedom and control contrasted sharply with his.

When Greg went back to the school after the fair, he tried to get used to, and adapt himself to, the extent of freedom the children were apparently used to having. He thought that perhaps by doing so, he could learn a lot from the experience of the people who had been at the SAU for some time. Besides, he thought that he had no other choice but to adapt, since he had been hired as an assistant teacher (he became the main teacher only when Amy went on leave to have her baby) and since he was on probation for a period of one month.

The more Greg tried to adapt himself to the freedom the SAU students demanded, the less comfortable he felt with the situation. Some of the students would make choices that, from his viewpoint, were not healthy for themselves, for their peers, or for the school as a whole. As he taught, Greg worried that some of the students—especially the older ones—would constantly resist engaging in activities proposed by the teachers, and would behave in ways that disturbed the children who wanted to work. He also worried that the students got engaged in certain kinds of "processings," which from his viewpoint, took too much time away from other activities.

> … I think that they waste a lot of time knowingly … in a calculated fashion to manipulate the adults to get out of work. I mean the time spent in meetings, in processing, little things that shouldn't be discussed. Teachers should be an authority and say no. And there is a point where processes—solely the processes—should not outweigh the product of what we are doing.…There's value in this processing and in giving them that time to speak, think, process ideas, and compare challenges. But I think it should be done more in an academic context. Certainly, there are social things that we can all discuss in processings, but, not every little thing. We can waste the whole day trying to satisfy them. They get bored. They get restless in those meetings, but, at the same time, part of them wants them and says "oh … better than other work.

While Greg believed that children learned a lot while participating in different kinds of decision-making processes, including those about their own learning and behavior, he thought that the SAU children were allowed to do this to too great an extent. They often spent so much time and energy on certain issues that there was no time left for other activities. From his perspective, many of the SAU students weren't ready to handle the degree of freedom they had at the school.

As Greg realized that the children refused to engage in what he considered important learning experiences, and after having to face many situations in which children kept hurting themselves and other children, Greg decided to take more control over the students' experiences at the school. But the more Greg tried to structure his classes and become directive, the more the students misbehaved and resisted him. This led him to constantly try different strategies that varied from being very structured to completely loose, according to what he perceived children needed. But while these variations were meant to help the students, they apparently confused them. In time, the students felt they no longer knew exactly what Greg expected from them, and how he would react to their work and behavior.

Let me give an example of how Greg's inconsistency might have confused the students. On April 9, 1998, I observed a class in which he read and discussed an article with the older group of students. For most of this class, he didn't seem to mind that, as the students were reading, and asking and answering questions, they would stand up, sit down, lie on the ground, sit up again, go to the bathroom, etc. Once during this class, two of the students actually ran outside to get something they had thrown through the open window, and Greg didn't seem to mind. However, after about 40 minutes of reading, discussion, and all this movement on the part of the students, Greg unexpectedly gave two of the children a warning because they were gently tapping against each other's knees. The students got so upset by the warning that, from that moment on, they started purposely distracting everyone else and stopped participating in the discussion. It is likely that the reason they got so upset was that in their mind, Greg had been arbitrary in his decision to give them a warning. How come he had allowed everyone to move around the whole class and then suddenly gave them a warning just because they were tapping against each other's knees?

Greg believes that most of his problems with the students happened because he started the year with the false expectation that the students were more ready than they actually were to handle the degree of freedom they demanded. He regretted that he and the other teachers, instead of having the students earn their own freedom, granted it to them for free.

From Greg's perspective, before children can have the freedom to make certain kinds of decisions, they need to demonstrate that they can make good choices and can handle all the responsibilities that freedom entails. Also, from his perspective, a possible strategy to help students develop responsibility for their own learning and behavior, is the implementation of some kind of structure that leads them to systematically process their experiences on a daily basis.

AMY

Amy believes that, since teachers are more experienced than students, they should make decisions about what kinds of things children should experience while at school. But as the teachers make such decisions, they should also take into consideration the interests of the children with whom they are working and should develop pedagogical approaches that are appealing to them. From Amy's perspective, the education of children should be a result of a compromise between what the teacher knows the child needs to know, and what the child is interested in.

> I would like to have something like a give and take: from the interests of the students to what I see as an adult—from my life experiences—the pieces that they might need ... so it is not an external piece of education being thrown at them, but is ... coming from them, and it is coming from who they are, and then the external piece [my input] is interwoven with that, like a combination.

Amy thinks that too much structure and control over students' experiences at school might be detrimental to them in the sense that this might prevent them from being able to be who they are and to pursue their interests. On the other hand, too much freedom might also be detrimental to them, since children are inexperienced and do not always know how to make the right choices for themselves: They do not know what skills and knowledge they might need in the long run so that they can become independent and life-long learners.

Like Greg, one of Amy's main concerns as a SAU teacher was that not all students were ready to handle the degree of freedom they were allowed at the school. While some of them did well with it, others needed a lot more structure so they could overcome some of their difficulties and keep working and learning. She was also concerned that perhaps one of the reasons some of the children were not thriving academically was that they had some kind of undetected learning disability. Since they had a lot of freedom to do what they wanted, they might have kept choosing to do only the kinds of things with which they felt comfortable, avoiding areas in which they were weak, making it difficult for the teachers to detect and address their learning challenges.[1]

Amy set quality standards for the students' work and tried to structure their experiences in ways that would help them achieve these standards. This worked well for some of the students, especially for the younger ones. The older students, however, kept on trying to find strategies to get out of doing what they were supposed to do, as they did with the other two teachers. But as Amy was only part time at the school, and as she didn't spend a lot of time with the older students (Greg had the older ones for most of the time during the development of the theme project), she didn't have to face nearly as many difficulties as Greg did to have them engage in academic work.

Amy was energetic and optimistic with the students: Energetic not only in the sense that she strongly encouraged them to work, but also that she was so confident they could do what was expected of them, and so enthusiastic about what they did do during her classes, that she sometimes stimulated everybody to work more and better than they initially wanted to. At the same time Amy was energetic and optimistic, she was also firm but gentle with the students. See, for instance, the following passage from a school meeting in which she, Greg, and the students were planning for the Acadia trip:

Claus - I have a problem with burgers and hot dogs!!! [Claus has a loud and aggressive tone of voice]

Amy – Why?

Claus - I am a vegetarian!

Amy - What do you eat at cookouts?

Claus – Nothing. I don't like burned food !

Amy – So what do you want to eat? You need to think about it and discuss it with the Acadia group!

Claus – I also have a problem with the tents!!! When we went to the Common Ground Fair, the six-man tent barely fit five people!!! [He is still speaking loudly and aggressively]

Amy – OK, so we'll limit the number of people!

Nathaniel – Yes, but the campsite might fit no more than four tents [uses aggressive tone of voice again]

Amy – OK, I'll call the park and see what they have there [Amy's voice is calm and soft throughout the whole conversation]

In the situation above, Amy was firm as she kept to the agenda of the meeting and did not let the students get lost in a discussion of small details that could be easily addressed outside the meeting. She also remained calm and was gentle as she addressed the students' concerns, even when the latter had quite an aggressive tone of voice.

Amy also had other characteristics that facilitated her teaching and relationship with students as she dealt with their resistance to listening to her and to the other teachers. She was genuinely and deeply open to listening to the students. Before trying to understand and analyze what they would say to her, she would usually listen to them closely, which helped her hear beyond her own "voice," and allowed her to capture more precisely what the students had to say to her. This gave her a degree of awareness of what was going on with the students, which helped her to effectively prevent and mediate problems she had or could possibly have with them.

Amy was also a skilled discussion leader. As she talked to the students, she would let them know she was listening to them and trying to understand their concerns and feelings. She would also ask them questions to clarify their thoughts and would summarize their points to make sure that she and everyone else had understood them well. Besides, she would acknowledge the positive aspects of their comments before she made remarks on the weaknesses in their arguments, and before she took a stand on the issue under debate. Because Amy would take all these steps before she passed judgment on students' viewpoints and demands, in most circumstances, students would end their conversations with her feeling they had been heard and that their viewpoints had been considered. Therefore, they would usually feel her decisions were fair, even when these decisions went against their personal positions.

When Amy related to students, she usually assumed that they were good inside, and that if they had done something bad, it had not been on purpose. Amy's qualities helped her prevent and mediate ongoing conflicts she had with the students as she tried to structure their experiences and have them do their academic work at the school. During one of my interviews with Amy, I asked her what she would do if her students decided they didn't want to learn the things that she considered important. Her answer was that she would consider a series of factors: First, she would analyze whether the things she wanted them to learn were, in fact, important. If she found they were not, she would probably give them up. If she found that they were important, she would try to show students how these things could be useful to them. Second, she would invite the students to try to discover with her ways in which they could go about those things that would work for them. Third, she would take into consideration the reasons the students resisted learning. It could be fear and insecurity that she needed to carefully address with them.

Both Amy and Greg actually did follow the steps Amy mentioned above. Before implementing any new learning unit, they would talk with the students about it and explain the relevance of what they were about to learn. If the students objected too strongly, they would not force it upon them, and would try to find other ways of working toward the same goals through different activities. As they taught, they also took the children's suggestions into consideration. Besides, they worked individually with each student in order to address his/her specific interests, needs, and fears. However, in spite of all this, some of the students would still resist the idea of engaging in most academic activities proposed to them. Or, when they did engage in these activities, they tended to give them up before the completion of their projects.

Comparing Amy, Kendall and Greg

Amy's and Greg's understanding of how much freedom should be granted to students sharply contrasts with Kendall's. While Kendall believes in granting children freedom to set their own goals and to pursue them, Amy and Greg believe that teachers, because they are more mature and experienced than the children, should carefully guide their learning toward goals they know are important. For them the curriculum, while it should be centered on the child, should not be a "free-flowing thing." It should be structured in ways that are conducive to growth and development, as Dewey (1963) remarked in his book *Experience and*

Education (Dewey was one of the authors who had the most influence on Amy's philosophy of education):

> The greater maturity of experience which should belong to the adult as educator puts him in a position to evaluate each experience of the young in a way in which the one having the less mature experience cannot do. It is then the business of the educator to see in what direction an experience is heading. There is no point in his being more mature if, instead of using his greater insight to help organize the conditions of the experience of the immature, he throws away his insight ... he must, if he is an educator, be able to judge what attitudes are actually conducive to continued growth and what are detrimental. (Dewey, 1963, p. 38)

Both Greg and Amy believe that children should only be granted freedom to make important decisions about their own education if they can demonstrate they can make intelligent decisions about their learning and behavior. As both of them were convinced that many of the SAU students needed a lot of guidance, they tried to provide them with a certain degree of structure. But, as they went about providing the kind of structure they thought the students needed, they had completely different experiences.

Amy was familiar with the school and with the kind of students who attended. She had not only attended the SAU up to the eighth grade, but she had been raised and educated at home by parents who fully embraced the philosophy of the school. Greg, unlike Amy, had had a traditional educational background both at home and at school. He wasn't used to alternative educational settings or to the degree of freedom the SAU students demanded. Therefore, it was more difficult for him to adapt to the reality of the SAU and to deal with the kinds of problems inherent in that reality. While Amy seemed to be at ease with the school, with the students, and with the problems that emerged during the year, Greg was uneasy and had a more difficult relationship with the students.

From my perspective Greg had a difficult relationship with the students not only because he didn't have much experience with alternative education—and this kind of education didn't seem to came naturally for him—but also because of some of the inconsistencies of his practice. As I have previously remarked, at times he gave students freedom, and at times he tried to be directive. This obviously led students to resist his way of being. Another factor that seemed to aggravate his relationship with students was the inconsistency that existed between his practice and that of Kendall's. To clarify the extent to which their practices differed and how this probably had an impact on students, I would like to offer a concrete example:

On a given day, I followed the older group of students both in Kendall's and Greg's class:

> It is about 9:30 and Kendall starts reading a chapter of a book to her students. About 5 minutes later, one of the students screams: "SKUNK!" and everyone runs outside to look for the animal. After a while, they ask Kendall whether they could do some free writing at Crystal Falls, a spot in the woods by a stream. Kendall agrees, and when they get to Crystal Falls, while some of the children play in the stream, Kendall and Karl start a campfire. She then asks the children to sit around the fire and do some writing. They write for no more than 2 minutes and claim they are done for the day. Kendall doesn't complain and she and one of the students spend the rest of the class time talking with each other, while the other children play at the stream or cook bread and hot dogs in the fire. After Kendall's class, the students go back to the school building to have a class with Greg. The latter asks them to read and discuss an article about four pages long written in small letters. After reading the article, which takes the children about 20 minutes, Greg requires them to participate in a focused discussion about the reading for about 40 minutes.

As we can see above, the difference between Kendall's and Greg's work requirements was striking. The students obviously noticed and reacted to them. After being allowed to do basically what they wanted during Kendall's classes, they naturally resisted the level of difficulty and the amount of work Greg required of them. Besides, after being with Kendall, who took her time to have fun with them (e.g., the hike in the woods, the campfire, and the play and the conversations that went on with her), the students naturally resented the fact that Greg was so focused on academic work. Even though, in the eyes of an outsider, Greg was relatively liberal and friendly with students, for the students who lived the contrast between his pedagogy and Kendall's, he seemed authoritarian and out of touch with them.

But, while the students realized the differences between the teachers and capitalized on these differences to resist engaging in academic work during Greg's classes, they were concerned about the degree of freedom they had in Kendall's classes. While on one hand, they were happy with this freedom, on the other hand, they were worried about it. Many of the students, even though they loved Kendall's classes, knew they needed to work harder in order to learn the kinds of things other children of their age are supposed to know. Below is a little piece of conversation I had with Abiel, where he expressed this concern as he answered my question "Would you change anything, and what would you change, if you could change things at the school?"

> Abiel – We could have more writing time. I don't feel as if we are writing a lot.
>
> Dora – And wouldn't you complain?
>
> Abiel – Probably, but I think it would be good for us.
>
> Dora – Why do you think it would be good?
>
> Abiel – Because I know we are not all that good in writing.... And I think we need to do spelling because I know I am really, really bad in spelling.

We continued this conversation on a different occasion when I asked him the following question:

> Dora – ... But at the same time you would like more structure, you complain when the teachers ask you to work. Why is that?
>
> Abiel – Ummm (silence). I guess because *I don't really want to do it, but I know I need to.* [my emphasis]

Like Abiel, other students—Claus, Kyle, Shandy, and Nathaniel—also expressed the idea that, even though they sometimes complained and tried to get away from doing the academic work the teachers asked them to, they knew that at some point they would have to work hard if they were to learn enough to be able to further pursue their education and dreams. When I asked the students whether they thought the school needed changes, and what kinds of changes they would implement, they made interesting comments relating to this issue. Allyson, for instance, said she would like Kendall to have them do more reading and spelling, and Claus and Karl said they would like Kendall to have them write more. Karl actually once directly asked Kendall to have them write more than they usually wrote when he overheard her telling me during one interview that she wished the children would write more in her classes.

> Just be strict ... you should be strict.... You should make us write more than five sentences.... You should actually maybe, bring towels and sit down at the beach and do a whole lesson there. I mean, I would enjoy that, and I would understand why you would do that.... I'll do it. It's only fair. I'm getting something out of it too! And then give us five minutes to go up and down the beach.

Shandy said he would like Kendall to give students more direction on what to do:

> Don't ask them what they want to do, you just say that you've got choices and these are your three choices and there are no more, and this is all I can do for you today, and maybe tomorrow you can do something funnier.

During my conversations with students, even Nathaniel, the student who most resisted engaging in academic activities, recognized he had to work harder on his academic development. He commented that the next year he would try to be a better student so that he could "get ready for high school."

There is one more piece of dialogue I would like to quote here because it elucidates the students' views on the extent of freedom and structure they had at the school. This conversation was also in response to the question "Would you change anything, and what would you change, if you could change things at the school?" When I asked Shandy this question, among other things, he said that he would like both Greg and Kendall to find a better balance between freedom and structure:

> Shandy – I like that we get to choose a lot of things [in Kendall's classes], but I would like a little more structure, because I feel

like when I go into Kendall's class, I have nothing to do, she is like oh yes, choose something to do.... Greg doesn't really understand the freedom of the school because he has been in public school the whole time....

Dora – At the same time, you said you wanted a little more structure from Kendall's class, you think Greg imposes too much structure?

Shandy – I want it to even up. Kendall is really nice and gives us a lot of freedom ... and Greg is totally the opposite.

Dora – So he is totally the opposite? So you want something in the middle?

Shandy – Yes, Amy is like right in the middle.

For many of the students Amy represented the right balance between freedom and structure: Something right in the middle between Kendall, who they thought asked too little from them, and Greg, who they thought tried to control them too much. It is interesting that the students perceived Amy as the teacher who best balanced freedom and structure, considering that her position on this issue was very similar to Greg's. I believe several factors played an important role in students' preference for Amy: First, Amy's interpersonal qualities (which I've already discussed) which helped her prevent and mediate possible problems she had or could possibly have with the children. Second, as Amy herself said, Greg was more willing than she to take on the "bad guy role." This obviously led the students to develop reservations about him. Third, since Amy taught only part time, and was not responsible for any of the core disciplines (math or language arts), she could be more flexible in terms of behavioral and work requirements, which helped her prevent power struggles with students.

Some students needed more guidance than others in terms of what they and their peers could and could not do at the school. For instance, James was constantly asking the teachers questions such as "Can students eat candy during snack time?" or "Can students play cards during lunch?" "Is everybody supposed to be going out right now?" Apparently, in order to be comfortable at the school, James needed the teachers to enforce certain rules. Like James, Allyson too demonstrated she would like to see a little bit more structure at the school, especially in what was related to behavioral issues. She said she wished the teachers made students "do what they are supposed to do instead of just wandering around in the school." She also revealed that she felt constantly harassed by one of the students who would often make rude comments to her and that she wished the teachers "would not put up with this student's kind of behavior." While both James and Allyson were excellent students academically, and did not need anyone controlling them so that they would do their work, they clearly needed some kind of structure in order to feel safer and more comfortable at the SAU.

When Claus and Abiel talked about changes at the school, they said they would like to have a larger amount of structured academic disciplines in the school schedule. Instead of having math and language arts every day, they would like to have more of other disciplines such as geography, history, sciences and physical education. According to them, even though they really appreciated having a lot of free time at the school, they would rather have a more structured schedule that made them work more. Below is a bit of conversation I had with Claus. I believe this quote is perfect to close this conversation about freedom and structure, since he brings up not only these points, but many others I have raised throughout this chapter.

Dora – Do you feel as if you are learning enough here at the school?

Claus – Yes … but I would like to learn more.

Dora – And what would you like to learn more about?

Claus – I'm not sure, but I would like to have more classes like more history, geography, and stuff like that.

Dora – … How much free time and how much academic work would you have if you could decide things here?

Claus – I'm not sure about that … like we would have the same amount of free time, but we could stick to a better schedule instead of going 20 minutes over on free time, coming in right at the end of free time and having, one day, having more language arts and math and the other like language arts and social studies and on the other day, having geography and math and something like that.

Dora – So like the days that were language arts days, like the days you went to the beach....

Claus – Yes, I don't think that really worked because we did more play than we really worked.

Dora – I see....

Claus – So we basically got there, had the morning meeting, we would go to the beach, have fun there, come back, have a snack and then have math and then free time and then that was basically the day.

Dora – So you feel that it could have had more structure?

Claus – Yes, more structure.

Dora – ... But do you feel that sometimes Greg is trying to do what you are saying, to have more academic work?

Claus – Yes, but it gets kind of boring because, it would be a lot more fun if we didn't have math and language arts every day, like if we had math and language arts one day, like I said, and then a schedule of different kinds of things on different days. If we had math twice a week and not every day.

Dora – I see.

Claus – Instead of math and language every day we could have different classes, geography and ... other days

D – Why do you think it would be better?

Claus – Because I think it would be less boring. More learning and less boring.

Dora – Why do you think this way is boring?

Claus – Because it's the same thing over and over every day!

Dora – But Wednesdays and Fridays are different....

Claus – But they are the same ... they are basically free days. There are only one or two things we need to do. I think if we could actually do academic work on Wednesdays and just have sports and free days on Fridays, I think that would work.

Dora – Speaking of sports, would you like to have more sports too or not? I noticed that you do a lot of movement but not structured sports....

Claus – Yes, more sports like phys-ed kind of sports.

Dora – Why do you think this would be important?

Claus – Uh, because it would give us a break from the same everyday thing. We would be able to learn some new sports and have fun while we are learning them ... we could have a team like a soccer team or a basketball thing.

As we can see in the dialogue above, Claus was a little bored by having the same academic disciplines every day. He was also bored because he was not asked to work harder at the school. It is interesting that while Claus, Abiel and other students said they wanted the teachers to be more structured in terms of schedule and requirements, whenever the teachers tried to have them in class on time and have them work harder, they complained about it and resisted doing what the teachers asked. Even though these students, in theory, were aware of the paradox between their "wants" (to play) and their "needs" (to work harder), and wanted the teachers to help them meet their needs, in practice, they would almost always prefer to go for their "wants."

Amy, Kendall, and Greg wished that their students would work more than they actually did, and they all tried to implement some kind of structure to help the children progress in their academic development. Note that, even though Kendall didn't structure her classes to the same extent as the other teachers, her classes did have a defined framework (e.g., in each of her classes, she would always have the students write, and she would either read to them or have them read by themselves, even if just for a little while). In spite of the degree of freedom or structure that each of the SAU teachers offered the students, none of them were able to change the reality that many of the students would resist engaging in academic activities. As I finish this chapter, I am left with difficult questions I would like to address in a future research project:

Is students' resistance to academic learning a central problem in other schools that give students a large degree of freedom to make decisions about their own education? How do some of these schools deal with this issue? Are the teachers in these schools able to scaffold students' interests into rich learning experiences? How?

Note

1. This kind of problem is in fact a typical issue at progressive schools.

Chapter 5

Learning from the SAU Experience

Issues and Recommendations

In this chapter, I will address some of the main difficulties experienced by the SAU community and elaborate on some suggestions that could possibly help both the school and those of us interested in holistic approaches to education to further develop the theory and practice of these approaches.

The School Community Needs to Develop and Commit to a Common Vision of the School's Educational Philosophy

One of the main problems of the SAU community is that it doesn't have a common vision of what the school is. The educational philosophy of the school is stated in broad terms in the school handbook and in other school documents. For instance, according to the school handbook, one of the main goals of the school is to help children "learn how to learn for life," and one of the ways through which this goal can be achieved is "by respecting choice as central to learning" (Board of Trustees of the School Around Us, 1997). However, in the handbook, there isn't a clear position on what exactly the SAU community means by "respecting choice" in the daily life of the school. Should children be allowed to make choices in any circumstances or should they do so only on certain occasions and within a well-defined structure? If the latter is the case, within what kind of structure should these choices be made? Considering the school doesn't have someone (e.g., a director or a coordinator) directly responsible for helping people clarify its educational philosophy and for answering questions such as those above, as people read and talk about the school's philosophy, they tend to understand and think about it in different ways. While this allows the school community to be open and adaptable to the changing and emerging needs of the school population, it also leads to problems.

From my viewpoint, the first step the school needs to address as it undertakes the task of clarifying its philosophy of education is to take a stand on the issue of freedom and structure and create clear policy to deal with it. This is important so that parents, teachers and students will better know what is ahead of them before enrolling in the school. Clearly some students started the SAU thinking that they could do whatever they wanted in that environment. Greg, on the other hand, thought the school was much more structured than it actually was. Had he known in advance where he was stepping in, he would probably not have applied for a teaching position there. The lack of definition on the issue of freedom and structure, as I have previously discussed, generates inconsistencies which may impair both the teachers' work and children's learning and development.

As the educational philosophy of the SAU is better defined, the school will probably need to implement a more thoughtful admission process to make sure prospective students and parents know what the school is about and whether they are ready to commit to take on the responsibilities that being a member of the SAU community entails. Prospective students should know, for instance, that freedom requires certain responsibilities. Parents should be conscious that, as Stacy Wentworth (one of the founders of the school) remarked during an interview, the SAU requires their involvement not only with the administration of the school, but also in the development of the curriculum. This is important if the school is to preserve one of the main concepts that led to its foundation, which is to have the whole school community—and also members outside the school community—fully involved with the education of children. This is also important if parents are to provide continuity between the work done at school and what happens at home.

Creating a Structure that Provides for the Balanced Development of All Aspects of the Child

Many of us have tendencies to focus more on certain aspects of our being—and other people's being—than on others. Kendall, for instance, tended to focus more on the spiritual and emotional self of the children than on their intellectual self. For her, more important than developing reading and writing skills was that students actually learned to enjoy reading and writing. Greg, on the other hand, even though he consciously tried to pay attention to the whole self of the children, was apparently more focused on their intellectual development than on their spiritual or emotional aspects (maybe this was his way of compensating for both Kendall's and the children's tendencies to focus on the other aspects). The children themselves tended to pay more attention to the needs of their physical self than to the development of their intellectual self. When allowed to choose their activities, they would almost always prefer to play physical games (e.g., basketball, soccer, or tetherball) than to work on academic projects. Considering that we all have different tendencies, if we are to provide a truly comprehensive holistic education, we need to devise some kind of structure that can help us provide for the balanced development of all aspects of children. Otherwise, we might get stuck in promoting only one or a few of their aspects according to our own or to the children's immediate interests.

Providing a Structure for the Development of Academic Skills

While giving freedom to children to make choices about their own education might enrich their educational experiences in many ways, it might also generate problems. As we could see by the experience of some of the SAU students, children might make choices that lead them to miss valuable learning experiences. For example, during the development of the "theme project," some of the students chose to focus only on the practical aspects of their work: the building of an obstacle course and of a golf course. They did so for almost seven weeks, in spite of all the teachers' efforts to have them focus on the academic piece of their projects. It was only in the last week of the project that, under a lot of pressure from teachers, they worked on the academic piece that was required of them.

Children are only children. They lack experience to know what might be important for them to know in the future. What if, in order to be free and happy then, they need academic skills they never developed in school? What if, when they find out they need these skills, they will no longer have conditions (e.g., time, energy, economic stability) to invest themselves in the kind of learning that would lead them to these skills? From my viewpoint, it is important that teachers provide students with a degree of structure that guides them toward the acquisition of skills they are likely to need to cope well in the world outside school and to further pursue their education. They will not be truly free in their adulthood if they do not develop the tools they are likely to need to meet their personal and professional demands.

Developing Strategies to Help Students Focus and Attain Depth in their Learning Experiences

Some of the characteristics of the SAU were both positive and troublesome. For instance, the freedom that the school provided the students to freely move around and to talk about issues that concerned them was positive in the sense that this allowed them to be themselves, to express their thoughts and feelings, and to let their physical energy out. But by the same token, this same freedom also allowed them to develop a certain kind of attitude that fostered disruption and that often prevented them from being able to focus and to reach a deeper level of involvement in their academic work.

Another characteristic of the school that had both a positive and a problematic side was related to what I came to call the "pedagogy of connecting with the real world." This pedagogy was positive in the sense that it promoted meaningful learning. However, it was problematic in the sense that the students sometimes got too overwhelmed with the plethora of learning possibilities offered by the real world. While connecting with these numerous and always emerging learning opportunities, many of the students got disconnected from the learning in which they had previously decided to involve themselves. And while moving from one experience to another, they often were not able to focus and to get in depth into specific experiences.

The other characteristic of the school that was positive but also troublesome, was what I came to call the "pedagogy of connecting the curriculum with children's emergent interests." This pedagogy was wonderful in terms of getting students interested and motivated to learn. However, because many of the children constantly changed their interests, they had a hard time attaining depth in their learning experiences and finishing their projects.

If both the "pedagogy of connecting with the real world" and the "pedagogy of connecting the curriculum with the emergent interests of the child" are to work, the SAU needs to figure out some strategies so that these pedagogies do not interfere with students' ability to focus, to get deeply into their learning experiences, and to complete their projects to their best ability, so that they can move beyond where they were before they started their learning activities.

Providing a Structure to Support the Students Who Need Extra Help and Guidance

Whereas some students can do very well if left free to make most choices about their own education, other students need extra help and guidance. Allyson, Elise, and James, for instance, took the freedom they were granted in Kendall's classes very seriously and worked hard to complete their projects to their best ability. But many of the other students, when given a choice, would almost always prefer to play rather than work, even when, deep in their hearts, they knew they had to work hard to develop certain skills they wanted and needed to have. As a consequence, they didn't intellectually challenge themselves and were not intellectually challenged as much as they could have been and as they sometimes wanted. These students definitely needed more guidance in their choices so they could achieve certain academic goals.

While some students need more help in the intellectual realm, others need more support in the social and emotional areas. Allyson and James, for example, had a hard time speaking up and making their voices heard in the large group. They also faced difficulties being included by their peers in group activities. Some kind of structure definitely needed to be created in order to help these students feel motivated and safe to express themselves, to engage in group conversations and decision-making processes, and/or to address the kinds of problems that prevent them from being well-received by their peers in group play and work.

Devising a System to Help Teachers Give a Fair Amount of Attention to all Students and to Keep Close Track of their Work

As the students had different interests, they pulled the teachers in different directions. Those who pulled harder tended to receive more attention than the others. During school meetings the older boys were usually more vocal. They often spoke about the subjects of their interest with so much passion that they ended up absorbing most of the teachers' and their peers' attention. These same older boys also ended up occupying most of the teachers' attention because they misbehaved a lot more than the younger students. The teachers often had to address the problems they caused before they could actually focus on their teaching and interactions with the other students.

Giving a fair amount of attention to all students is apparently quite difficult when different teachers work with the same group of students. This was the case of Amy and Greg, who taught together during the afternoon classes while the students developed the theme projects. As both these teachers worked with all the students, and as the students worked at different places in the school (outdoors and in different rooms indoors), they sometimes lost track of what each student was doing, and didn't know who had already received attention from the other teacher. As a consequence, on the same day, in some cases, a few of the students ended up receiving attention from both teachers, while others received no attention at all from either of them. Also, on some occasions, the same student received exactly the same kind of input and feedback he/she had already received from the other teacher.

During the academic year of 1998-1999, the school divided the group of students into three main groups: kindergarten, primary school and middle school. As the students stayed with a peer group of a similar age in the same group, and as each group had only one teacher, the problems mentioned above were addressed, at least to a certain extent. But still, if teachers are to be able to be more equitable in terms of sharing their time and energy and be more efficient in the use of their time with individual students, they need to develop some kind of system to monitor their interactions with individual children.

Fostering Systematic Processing on the Work Done at the School

From Greg's perspective, places like the SAU—schools that provide students with freedom and decision-making power over their own education—can only be successful to the extent that they require students to do some type of processing of their work on a regular basis. This would help students to reflect more on their own decisions and actions, and have them become more accountable for their learning and behavior.

As I observed the SAU students, I noticed that they missed opportunities to learn from their experiences because they didn't systematically reflect on their experiences. For example, on April 5, we went on a field trip to Portland. The children were divided into two groups and each of these groups was supposed to answer two different sequences of questions and have quizzes on the architecture, historical facts, and general information about the town. When we came back to the school, the answers to the questions were not discussed and the students never got together as a group to share what each child had learned during the trip. Perhaps the intention of the trip was recreational, and this is why the teachers did not facilitate activities related to the trip after it happened. However, since the two groups of students had visited different parts of the town and answered different questions and quizzes, if they had gotten together to share their experiences, they would probably have learned different things from each other.

While both teachers and students seemed to learn many things during the SAU field trips and other SAU activities, their learning experiences could probably have been even richer had they developed a system to discuss and process their experiences as a group. The processing of experiences would also be valuable in the sense of helping the children develop a habit of constantly reflecting on everything they did, so that all their experiences could turn into learning experiences.

Defining a Clear, Consistent, and Efficient Discipline Policy and Establishing a Comprehensive and Detailed Evaluation System

The lack of a well-defined and consistent structure to deal with behavioral issues from the beginning of the year led the SAU teachers to spend an inordinate amount of time dealing with behavioral problems. As Greg himself said, had he and the other SAU teachers developed a stronger behavioral policy right at the beginning of the year, they would probably have prevented many of the behavioral problems they had to deal with as they tried to work with the students.

Whereas the SAU evaluation system was comprehensive and talked about all aspects of the child (the teachers wrote a letter about the development of each student), it lacked detailed information on how each child was doing in each specific area. For example, in these letters, the teachers would make broad statements such as "so and so has made significant progress in reading and writing." However, they would then not specify exactly in what areas of reading and writing the child had progressed.

Besides, Amy's, Kendall's and Greg's letter focused almost exclusively on students' strengths. While this fostered the latter's self-esteem and self-confidence, on the other hand, it did not induce them to face and work on some of their weaknesses. I saw this as a potential problem since sometimes, schools are the only place children can safely do this, and therefore, they just can't afford to miss the opportunity to confront and overcome these weaknesses while still at school. Even though this was probably not the case of the SAU students during the time frame I observed the school, this could be true in different circumstances.

The broad evaluation done by the SAU teachers—like grade evaluation in most traditional schools—did not tell the teacher, the student or the child's parents exactly what the child had learned and accomplished during the year, and in what areas she still needed to improve. A better evaluation system would consist of having periodic and systematic detailed reports describing the activities the child involved herself in, how she did in these activities, and what kinds of strategies the teacher used that seemed to best help the child learn. This kind of report, along with samples of the child's work could give the child, her parents, and the child's future teachers some concrete basis to think about what kinds of experiences the child had already had, what areas she needs to improve the most, and what kinds of strategies are likely to work to best help her learn.

The Beauty of the School Around Us

After pointing out so many problems and difficulties of the SAU, I would like to conclude my portrait of the school by summarizing some of its strengths so that we do not end this work with the false impression that the kind of education promoted by it doesn't work. Based on my observations, some of the most important strengths of the school are:

- Learning happens in the real world and is based on real-life events. This helps students see the practical application of everything they learn as they are learning, which makes learning more real and meaningful to them. Besides, this fosters in them an understanding that the world is a school from which they can always learn.
- The school is not focused exclusively on the intellectual self of the child, as is the case in most traditional schools. It respects and honors the child as a whole being and takes into consideration all her aspects: intellectual, emotional, physical, spiritual and social.
- Children are allowed to be children, to play, to run around, to appreciate the beauty around them, and to enjoy each moment they spend at the school. They are not rushed from one activity to the next, but are allowed to maintain their spontaneity, sense of wonder, and ability to appreciate their experiences as they happen.

- Children are encouraged to express their feelings, thoughts, and ideas. This not only gives them emotional support, but helps them maintain their sense of identity and self-worth.
- Students learn about, and learn to be a part of, democratic processes as they run school meetings and make group decisions about their experiences at the school.
- Students learn to work with others as they solve problems together both at school meetings and in learning activities.
- Students are encouraged to spend a lot of time playing and working outdoors, which promotes their sense of connection with and appreciation for the natural environment, and a sense of physical and spiritual wellness.
- Students develop a sense of social awareness and responsibility as they work on community service projects.
- Children learn about the importance of recycling and actually develop the habit of recycling materials.
- Children care about and care for each other. The older students take care of and teach their younger peers, and as they do so, they also learn. As the younger children receive attention from their elders, they benefit not only in the sense that they are cared for and learn new things, but in the sense that, as they are exposed to this kind of care and education, they too develop a caring attitude towards others.
- Children learn to respect and care for common spaces they share with people as they help in cleaning, maintaining and improving the school facilities.

Next, I would like to cite a few remarks the students themselves made about the school during my interviews with them that speak about some of the SAU strengths:

- "Learning and fun go together."
- "You can have a real childhood."
- "There are lots of hands-on activities."
- "You can learn at your own pace."
- "You learn for yourself and not for the teacher."

- "The school is like home: It is comfortable, I know everyone and everyone knows me."

Before concluding this section, I would also like to list some qualities that former SAU students claim to have developed at the SAU.[1] As these qualities stayed with them for a long period of time after they left the SAU, they also highlight many of the strengths of the SAU:

- Awareness of personal strengths and ability to work with these strengths to accomplish goals.
- Ability to maintain a sense of identity and purpose even in the face of adverse situations.
- Feeling of empowerment to do the things one wants.
- Respect and appreciation for the natural environment.
- Good communication and listening skills.
- Good conflict resolution skills.
- Ability to work cooperatively with others.
- Good problem solving skills.
- Ability to look at issues from different viewpoints and angles.
- Ability to participate in democratic decision-making processes.
- Desire to work with people of the community to accomplish communal goals.
- Desire to work with others to change things for the better.
- Good self-advocacy skills.

As we consider the gifts and limitations of the kind of education being offered by the SAU, we need to examine our deepest beliefs on what education is for and what an educated person means to us. Depending on our answer to these questions, we will probably look at the SAU experience with different eyes. An educated person for me is one who is self-disciplined and systematically strives to know oneself so that he/she can become better than he/she already is. He/she can live in harmony with the natural environment, with other people and with oneself. He/she is able to do the things he/she likes and would like to do, and while doing these things, make choices that are constructive to him/her and to other people's happiness and well being. In addition, an educated person is also one who knows deeply about and functions well in the world in which he/she lives. And, because he/she does so, is able to work toward transforming this world into a better place to be. For me, a truly educated person needs to have all the characteristics above, and not only a few of them. This assumption served as the lens through which I have looked at and examined the SAU experience—and that of the other schools I will be talking about in this book—throughout my data collection and analysis.

As I finish my report on the SAU, I am left with some difficult questions. Considering that the SAU experience tends to be conducive to the development of just some of the above mentioned characteristics, and considering that the experience of most traditional schools is usually more conducive to preparing students to function well in our fast-paced and competitive world, I wonder if it is really possible to educate people for the attainment of *all* the traits I refer to above in the context of schools. Are they compatible with each other? Is there enough time to foster all of them within the limited time frame of schools, since the school schedule is usually limited to six hours a day? If so, how is this to be done? Perhaps the research on the two other schools which I will present next will shed some light in these questions, but still, a lot more research needs to be done so that we can fully answer them.

NOTE

1. I drew this data from two different videos. One was produced by John Bordage and Caleb Clark (1989) as a response to a survey sent out by Marilyn Wentworth to people involved in the SAU (the goal of this survey was to investigate people's experiences and perspectives on the school). The other was produced by Claudia Berman (1994b), who interviewed a group of alumni and alumnae on "things that went well and not so well at the SAU."

Chapter 6

Feedback on Research Findings

As I noted in the methodology section, an important part of my research was to share my findings with the participants of my study to get feedback. Amy, Kendall, Greg, and Claudia Berman read my final research report (my dissertation, which was later transformed into a section of this work) and endorsed the internal validity of my research. This epilogue is dedicated to presenting and discussing the points they raised in response to my findings.

I will start with Greg's feedback. According to him, as he read my research report, he "felt all the joy and pain" that he felt when he taught at the school (telephone conversation, August 1999). He also said that he did not have much to comment on about what I had written since, from his viewpoint, I have addressed all the main points of his educational philosophy and practice, and have painted a portrait of the school that well represented what was going on there during the time I conducted my research. He did, however, bring up one main issue which I should mention here. He commented that I missed some interesting things he and the other teachers had done throughout the year, since I observed the school for only three months. I fully agree with him and this was certainly a limitation of my research. Nevertheless, because the data collection was quite intense in this period of three months, I was able to amass enough qualitative data to allow me to get in depth in many of the issues discussed throughout this work.

Kendall brought up three main issues as a response to my research report. She first reinforced the idea that, from her viewpoint, students were learning all the time, even when they did not seem to be learning. They learned as they played, they learned in the school van on their way to different places, they learned from the problems they faced, and they learned from the freedom they had in her classes. For instance, one important lesson they learned from their freedom was that they needed more structure:

> This is an important lesson. The children learned they wanted more structure. Would they have learned this without the freedom? This will help them take more responsibility for their own learning." (Handwritten notes on research report, September, 1999)

The second point that Kendall brought up was that children will learn whatever they need at any point in their lives. I agree with Kendall on both points. However, from my perspective, the degree of freedom promoted in the SAU is in many cases problematic for reasons which I will thoroughly discuss in the conclusion of this book. For now, I just would like to mention that schools that do not maintain a good and well-structured academic program might be a viable alternative only for families who can provide their children with extra help in the academic area and/or offer them some kind of economic safety net when they grow older. Children who do not have support for their academic development at home and who do not have a financial safety net to protect them later in their lives might find themselves trapped in a vicious cycle, which is typical among the poor. They might find themselves having to work a lot—sometimes in jobs they don't like and are not well paid for—because they do not have the basic skills that would allow them to work in more meaningful and well-paying professions. And they might also find that they cannot work on the development of skills to land better jobs because they are simply too busy trying to earn a living for themselves and their family.

The third issue Kendall brought up was that, while I did a good job in presenting the difficulties of the school, I did not talk much about the benefits of the kind of education offered by the SAU. Kendall is absolutely right. The focus of my research on the SAU was more on the problems than on the advantages of the education offered by this school. As I developed my dissertation, I had intentionally planned this focus because, while the holistic education literature talks about the rewards of holistic education, it hardly addresses its problems. Therefore I felt the problems needed to be detected and addressed if we were to—and still are to—in the long run, multiply the benefits of holistic education.

Amy hasn't had enough time to do a careful reading of my dissertation and to give me thorough feedback on my research, given her tight schedule as a mother and SAU teacher. Based on her quick reading of my work, she brought up two main points: First, she said I have developed a pretty accurate picture of her philosophy of education and of what was going on in the school during the time frame I was there. Second, the year I observed the school was a difficult year for the SAU, since both she and Greg were new teachers in the school, and since she worked only part time in the second semester and didn't have enough time to invest in the school. She suggested that, if we decide we want to develop a more thorough view of the SAU experience, we should probably come back to look at the school now that things have normalized and the school is structured in ways that are more conducive to its educational goals.

Claudia Berman brought up various issues as she gave me feedback on my research report. The first point she made was that, because the SAU curriculum is constantly emerging from the ongoing interests and needs of the school community, it is always changing. Therefore, since I observed the school for only a short period of time, I developed only a partial view of the SAU curriculum. Along the same lines as Amy, Claudia also emphatically remarked that I had observed the SAU in an unusual year, a year that was extremely problematic and that did not well represent the school. But from her perspective, while I developed a limited view of the SAU curriculum given the problems the school was facing the year I was there and the short time span of my research, I did capture a good view of the kinds of perspectives and challenges the SAU—and probably other holistic schools—often deal with.

> The issues that you raise are the issues of this kind of education, the issues that everybody is grappling with, and it might help the school a lot, especially to read the conclusion. (Audio-taped phone conversation, 9/8/99)

> After reading it [the dissertation] I was surprised at how well those three teachers, even though it was an unusual year ... actually represent the types of teachers that come to the school. They actually ended up representing the three different views, Amy is sort of in the middle, and Kendall and Greg on the other ends of the freedom and structure continuum. And they do represent what we usually get. (Audio-taped phone conversation, 9/2/99)

The second issue Claudia Berman brought up as feedback to my dissertation was that I should probably have reinforced more the idea that SAU offers holistic education not only for the students, but for their parents.

> It is phenomenal the way the school is run with this parent cooperative, where actually the whole parent group is going through holistic education by having to be part of the school. So we are actually training the whole community of people, including the adults, rather than just the students. And there are a lot of drawbacks, and a lot of problems; but then again, we are doing something a little different than a holistic school would which has a director. (Audio-taped phone conversation, 9/2/99)

I haven't discussed this aspect of the school, since the focus of my research was on the educational philosophy and practices of the teachers. However, Claudia is right when she says that the fact the SAU offers holistic education for the parents also is extremely important. This plays a strong role in what happens at the school and sharply distinguishes the SAU from other schools. After Berman made this point, it actually occurred to me that perhaps, inherent in the SAU understanding of holistic education is the idea that this kind of education should include not only the child, but also her parents. Maybe this is the reason the school community insists on not hiring a director and in keeping the parents running the school through consensual decision-making processes in spite of some of the problems these processes entail.

The parents' participation in the administration of the school and consensual processes lead the parents to constantly (1) evaluate their personal values and conceptions of education; (2) listen, consider, and honor the needs of other parents and children; (3) express their own needs and viewpoints; and (4) collaboratively problem-solve and work as a group. This leads them to learning, growth, and to "lessons of personal power" as Claudia (Berman, 1994a) says in her book. In other words, it leads them to the realization that they have an individual voice, a voice that can make a difference in their own and in other people's lives. The major drawback of this process is that, since every parent has an equal voice, and since the wants and needs of each SAU member should be taken into consideration before consensus can be reached, decision-making processes at the school are extremely slow. Therefore, any kind of improvement at the SAU usually takes much longer than it normally would at other schools with some kind of central administration.

The third issue that Claudia brought up is related to the issue of freedom and structure. From her viewpoint, it is healthy for the school to have teachers with a variety of perspectives, since this provides different learning opportunities for the students:

> If the students stay at the school for a long time, like my son and my daughter, they have had teachers from all places on that continuum of freedom and structure. They had teachers very free and "do what you want" and they had teachers that had structured the lessons and told them what to do. Kendall, for example, is a total gift for the school. She and her perspective and her way of being in the world are an amazing role model for people who may want that. So I guess, I love you came up with this word "pedagogy of connections." I feel like it is all like that. It is all connected ... and we don't have any one pedagogy in a way, we have a mixture and we honor that everyone has a gift and that this is valuable learning for our children to be in contact with all these gifts from all the different personalities." (Audio-taped phone conversation, 9/8/99)

She also remarked that perhaps it is healthy for the school not to clearly resolve the issue of freedom and structure. She believes that one of the advantages of the constant debate over this issue is that it allows for the expression and appreciation of individual viewpoints and for the adaptation of the school according to the needs of the school community. From her perspective, the resolution of the issue of freedom and/or structure and the adoption of pre-determined kinds of structure could possibly silence dissonant voices in the school community. This would go against one of the SAU's most basic principles; that is, that each individual person has his/her own wisdom and that "collective wisdom is wiser than the wisdom of one perspective" (e-mail message, 9/7/99).

I agree that the SAU openness to the constant debate on freedom and structure might be healthy for the reasons she pointed out. However, from my perspective, because the SAU population is highly volatile (as is the case with most schools in the modern world), this openness is more problematic than healthy. Since the school population is always changing, every new school member might raise the same issues discussed in previous years, and the same conversation might be carried on from year to year. As a consequence, the level of discussion of the SAU community might not progress far enough to start effectively addressing some of the problems discussed in the previous chapters.

From Donald Oliver's perspective (personal communication, September 10, 1999) the problems of the SAU as they relate to the issue of freedom and structure cannot be solved if the school is to maintain its identity and preserve its ability to fully take into consideration the individual voices and needs of each member of the school community. He believes that, in societies in which people live in the same place and attend the same school for a long time, people can have continuous and consistent dialogue and experiences within a less structured environment. However, in the modern world where people are always moving from place to place, and schools are seen only as transitory institutions, some contractual relationships are necessary to make the rules and to enforce these rules so that the school experience can have some continuity and consistency. From Oliver's perspective, schools like the SAU will only be able to maintain their identity—and at the same time address the kinds of problems I have raised throughout this dissertation—if we create a different kind of social system in which people live in one place for a long period of time and don't look at schools as transitory institutions.

Since it will take us a long time to create the kind of society Donald Oliver was talking about, it seems that for now, perhaps, a transitional alternative for the School Around Us is to develop some strong policies to try to keep the same teachers and families in the school for as long as possible. The parent group could try to keep their teachers by supporting their work more directly and by raising their salaries. I am aware that raising the teachers' salaries is a difficult task for the parents, since the school budget is tight. However, if this doesn't become a priority, good teachers will not stay long at the school, and it will probably never be able to keep a consistent program and will keep on facing many of the problems discussed in these chapters.

Another policy that could perhaps be implemented by the school is to prioritize the enrollment of families who will be committed to being a part of the school community for as long as possible. It is true that many families don't really know whether they will want to stay at the school before they experience being there. However, it is possible to predict—at least to a certain extent—whether the families are likely to fit in the SAU community through a detailed application process that includes various visits to the school, participation in parents' meetings, and detailed interviews with parents and students. According to Claudia Berman (audio-taped telephone conversation, 9/8/99), the school has become increasingly more selective in the admission of new families to the school since the year I developed my research.

It would be important to have a continuing group of families at the school for at least three basic reasons: First, members of the school could get to know each other well, establish common short- and long-term goals, and decide how they will work as a group to achieve these goals both in the short and in the long run. Second, the school community could maintain some sense of history and continuity, and avoid basic mistakes and problems which tend to repeat themselves from year to year. Third, the SAU community would become more cohesive in terms of its educational philosophy and practices. It is true that the SAU has already done an amazing job towards building community. But were the school population not so volatile, this community would probably be even stronger and more able to achieve an increasingly grounded and solid educational philosophy and practice.

The Sunshine Waldorf School

Chapter 7

Waldorf Education and the Sunshine Waldorf School

Waldorf Education

The Waldorf approach is an alternative kind of education started by Rudolf Steiner in Stuttgart, Germany in 1919 and today is a growing movement that includes approximately 750 schools in 44 countries around the world. Steiner (1861-1925) published over 50 books and gave about 6,000 lectures in different fields including philosophy, sciences, arts, and education. He started a movement called Anthroposophy, defined by him as "the wisdom of the human being" and as a "path of knowledge to guide the Spiritual in the human being to the Spiritual in the universe" (Trostli, 1998, p.1). The Anthroposophical movement is currently organized in many countries and offers insights to areas such as holistic medicine, holistic education, organic farming, and architecture.

Two of the most important themes of Steiner's work are spirituality and a critique of overconsumption (Henry, 1993). He makes a radical criticism of the materialistic worldview and universe of meaninglessness promoted by secular modern mentality, and proposes that instead of focusing on external achievements, we should work on our inner development and on a healthy and harmonious relationship with ourselves, with nature, and with other people and living things. He also criticizes the loss of spirituality in the modern world and proposes that scientific and spiritual insights go hand in hand for a true and complete picture of reality and for a healthier and more balanced life.

Steiner started the first Waldorf School at the request of Emil Molt, the owner and director of the Waldorf Astoria cigarette factory in Stuttgart right after the First World War. Those were difficult times for the German economy. Unemployment and social problems were rampant all around the country. In the face of this situation, Molt asked Steiner to idealize a knew kind of education and start a school for the children of his workers that could, in the long run, help promote some deep and positive changes in society. Steiner accepted the challenge and developed an educational philosophy and practice that profoundly contrasted—and still contrasts—with mainstream education. It focuses not only on children's intellectual development, but on the development of their whole being, and challenges the mechanistic, materialistic, consumerist, and fragmented worldview which permeates most educational practices.

Steiner's educational views and pedagogy rest on his notion of human development. He believed in life before and after death and in reincarnation. According to him, each person is "the expression of a divine spiritual being that descends from purely spirit-soul existence and evolves here in physical-bodily existence between birth and death" (cited in Marshak, 1997, p. 34). For him, the purpose of life is evolution and the attainment of divinity, which can be achieved as one works on his/her development in a succession of reincarnations.

According to Steiner, the child develops in stages of approximately seven years. The first stage lasts from birth to seven years old, the second from seven to fourteen, and the third from fourteen to twenty-one. During the first stage, he says, the child gradually comes down from the spiritual world into the body and tries to adapt to the use of the body as a vessel so that she can live on the earth. The life forces or energy of the child are focused on the growth and development of her physical body. Therefore, he claims, at this stage the curriculum should focus both on helping her adapt to her physical body and on fostering the development of her will.

Whereas during the first seven years of life most of the energies of the child are focused on physical development, during the second stage, according to Steiner, they are focused on the development of the child's "feeling life." Thus, he claims, the curriculum at this stage should center on nurturing the child's feeling life and imagination. During the third stage of development, from 14 to 21, the life forces center around the development of the child's intellect. It is only during this stage, and not before that, according to Steiner, that the curriculum should center on children's intellectual development. I will further discuss Steiner's views on human development and education in the chapters that follow.

THE SUNSHINE WALDORF SCHOOL (SWS)

The Sunshine Waldorf School[1] is a member of the Association of Waldorf Schools of North America (AWSNA), which means it closely follows the educational framework given by Rudolf Steiner. At the time I collected data for my research the school was almost thirty years old, and had a total of 316 students enrolled from nursery to 12th grade. The main purposes of the school—which reflect the purposes of all Waldorf education—are summarized as following in the school's handbook:

> In the largest sense, Waldorf education aims to establish balance in the lives of individuals in our world today. The school endeavors to awaken and to preserve the child's innate sense of wonder, awe, and reverence for life; to restore vitality to childhood by infusing the learning process with love and enthusiasm, and to cultivate the child's capacity for clear thinking. Our objective is to help all children blossom into young adulthood with a balanced capacity for both feeling and thinking so that they are prepared, with self-confidence and inner resources, to accept responsibility and take their places as creative, self-directed members of society.

Like other Waldorf schools, the SWS doesn't have a principal. It is run by the faculty, by a Board of Trustees, by a College of Teachers, and by a Steering Committee. The Board of Trustees includes parents, faculty and friends of the school and deals mostly with the economic health of the school. The College of Teachers includes a group of faculty who is committed to Anthroposophy and to the long-term development of the school. The College makes pedagogical and personnel decisions and is responsible for "holding the school on its course." The Steering Committee is formed by the faculty chair, the college chair, and the elementary and high school administrators. It is responsible for the connections between the school and the outside world. Both the College of Teachers and the Faculty meet once a week and the Board of Trustees once a month. All decisions at the College and the Faculty meetings are made by consensus. It is interesting to note here that the faculty meetings not only happen every week but also last for hours. During these meetings, members of the faculty discuss and make decisions about administrative issues and work on their personal and professional development. They also organize and participate in lectures and workshops on different topics and invest themselves in artistic activities such as paintings, drawing, and modeling with clay. Besides, they discuss issues directly related to their teaching, which gives them support to their self-development and practice.

The SWS is located in an affluent neighborhood next to conservation land, which is used as an extension of the school. The SWS facilities include playing fields, courts, a playground, an auditorium, a little gym, and a pond. The school building looks like a traditional school with long corridors and numerous classrooms. Within each classroom, the teacher's desk is placed in front, and students' desks are lined up in rows. The main difference between the SWS and most traditional schools in terms of physical appearance is that each classroom, according to the grade level, is painted with a specific color. Besides, the Waldorf School is beautifully and carefully decorated with children's artwork (the walls are neither empty nor overcrowded with children's work).

The SWS students come from different places. Many families actually moved to the town where the school is located or to neighboring towns in order to have their children in the school. At the time of my research, the tuition for elementary school was around $9,500 and the great majority of children were white and from middle and high-middle class background.

It is important to remark here, before the next chapter, that all Waldorf Schools—and I am talking here about Waldorf Schools which are recognized both by national and international Waldorf organizations as truly Waldorf—follow the principles and orientations given by Rudolf Steiner. Therefore, they look alike in many ways. For example, all Waldorf classrooms are painted with the very same colors according to each grade (each grade has a specific color), and the curriculum in all Waldorf Schools follows the same structure and thematic sequence. It is for this reason that at times, instead of speaking about the SWS specifically, I refer to Waldorf Schools in general. But when it comes to characteristics that might be unique to the SWS, such as, for instance, the quality of education offered by this institution and certain viewpoints expressed by the professionals who work there, I attribute these characteristics exclusively to this school and professionals.

Note

1. The Sunshine Waldorf School is a fictitious name and, if there is any school with that name somewhere in the world, it is a coincidence.

Chapter 8

A Curriculum of the Heart, Hand, and Head

There is a Right Time for Everything: Letting Children be Children

Waldorf educators are protective of childhood. They want children to be able to be children. According to them there is a right time for everything in life and childhood is a precious time that should not be rushed in the name of knowledge and skills that will supposedly prepare children for later life. They believe that when children are led to focus on intellectual development too early in their lives, they use up energy that should be used for other aspects of development. Consequently, they become deficient in these other aspects. Some of the teachers I interviewed in fact observed that some of their students who were "too advanced" intellectually and academically in the early grades, tended to lack the most in terms of emotional, physical and/or social development.

Waldorf educators believe they should foster children's "feeling life" before focusing on intellectual learning so that young people can develop a strong relationship with the world around them and a solid basis on which to grow academically. A teacher explains:

> First you build a healthy body and a rich soul life through hands-on experiences, drawing, painting, and music.... Then you have the foundation to develop the intellect.... Students come to me from public

> schools and they are filled with facts. And at the most fundamental level, I discover they have a weak relationship to the world because it does not come through their experience and their feeling life. Their imagination has not been given the time or the space to develop. It is just a careful tending.... If you plant a seed in rich soil, and you grow a really strong plant, that is the human being who can survive in the real world. If you take a seed and you put it in poor soil, and little sun and little rain, and tons of fertilizers, and expose it to real life circumstances, it is not going to survive.... (Samantha, 7th grade teacher)

In Waldorf schools the teaching of academic knowledge and skills is done later than it is usually done in most other schools. For example, they do not teach writing until first grade and they do not teach reading until second grade. Besides, they also usually do not encourage kindergarten and first grade students who come to school already reading and writing to further develop these skills before the whole class is ready for that. Instead, they reintroduce these children to reading and writing through the Waldorf method. Two positive aspects of this approach are: First, it provides children with time (and saves their energy) to work on aspects of their development that should be nourished while they are still young such as the development of their will, imagination, and gross coordination skills. Second, since it encompasses a lot of artistic activities such as drawing, painting, story telling, and short plays, it potentially "brings letters to life" and fosters a deep connection between the child and the written world. A negative aspect of this approach is that some of the students who already know how to read and write might feel held back. As I collected data in the second grade classroom, for example, I observed that while the teacher was still asking the class to read individual words from the board (initial stage of reading), some students would pull books up to 150 pages from under their desks to read when they had some free time.

Whereas I do agree with Waldorf teachers that children should not be forced into academic activities early in their lives, because we live in a culture in which children are exposed to the written world from the time when they are born (e.g., through ads, books, computers, and magazines), many of them are spontaneously learning to read and write at an early age. While we might want to redirect their interests into other kinds of learning, this is not always appropriate. Some children want to learn these skills so badly they just won't change their mind and/or would feel really frustrated if we tried to talk them out of it. There is also the case of children living in poverty who need basic reading, writing, and math skills early in life so they can survive in big cities. This is obviously not the case of the SWS students, and perhaps not the case of most children in the USA, but this certainly happens with millions of children living in other countries who need to fend for themselves in the streets (e.g., to take a bus they need to be able to read the name of the bus and to pay for their own fare). When this is the case, I believe, we do need to help children develop the skills and tools they need so that they can survive on their own.

The Waldorf Nursery and Kindergarten: A Brief Overview

As I have previously stated, the focus of my research is on elementary education. In this section I will offer just a brief overview of the Waldorf nursery and kindergarten to provide some background about what lies before the elementary grades.

When we step into a Waldorf nursery or kindergarten for the first time, we are likely to feel as if stepping into a different world: A surreal realm of fantasy, imagination, grace, and beauty. The classroom walls are colored with washes of pastel pink and muslin drapes hang from the ceiling softening the rectangular form of the room. The classroom is decorated with plants, flowers and other organic materials. There is virtually no plastic and most—if not all materials—are made of natural elements.

One cannot find manufactured toys in a Waldorf classroom. Students use pine cones, seashells, rocks, blossoms, tree branches, and pieces of wood , which they bring to life through their imagination. The few finished toys available in the classroom are all handmade and purposefully simple so that children can exercise their imaginative life. For example, all the dolls have a plain face so that children can imaginatively create the details of their facial expression and mentally determine the kind of mood they want the dolls to have for their play.

In Waldorf education, especially in nursery and kindergarten, the teachers try to make of each moment a special occasion and try to treat every single object as a special object almost as if everything were sacred. For example, if teachers and students are walking in the woods and if a child notices a flower, the teacher is likely to stop and appreciate the flower with her. Or if children make little objects, animals, or fairies out of beeswax, they will not just place them on a shelf, but will nest them in beautiful pieces of cloth so that everyone can admire them.

The development of children's sense of wonder is at the center of the curriculum. Once, for instance, I was observing the kindergarten classroom when suddenly a woman dressed up as Santa Lucia—she was wearing a long white gown, a hat decorated with flowers, leaves, and four lighted candles—came into the room singing a beautiful song. Yes, she had four lighted candles standing on her hat! She seemed to have jumped straight out of a fairytale book right into the real life of the classroom. Her presence was so impressive that everyone, including myself, was in awe. The beauty of that scene still lives in me very strongly as it probably still lives in many of the children who were a part of the episode.

At the nursery and kindergarten level there isn't formal instruction. Waldorf educators believe that young children learn best through free play, imitation and example. Therefore they see their main tasks as to offer children a beautiful and orderly environment and to be good role models for their students. They believe that children are "sense organs" that perceive and imitate not only people's actions and words, but also their inner attitudes. Therefore, they try to be as careful as possible in how they approach every little thing in their daily tasks. Trostli (1998) explains:

> Children also imitate our inner attitude. Kindergarten teachers therefore try to pervade everything they do with care. This will be reflected in the way they place an object on the seasonal table, or the way they put the toys away at clean-up time and make sure all the babies are tucked in and don't have any cold toes sticking out. If parents and teachers approach common life tasks such as cooking or cleaning with reverence and care, children will develop a deep respect for work and

> for material things. If, however, such tasks are done quickly and sloppily, this will be reflected in children's difficulties in finding meaning in life. (p. 93)

Some of the main activities I observed in the Waldorf nursery and kindergarten—which according to the literature is consistent with what happens in other Waldorf schools—were: free play (a lot of imaginative play), painting with primary colors, modeling with beeswax, singing, games with rhythmical movements, and outdoor walks. I also observed a lot of story telling (the kind of story telling that isn't read from a book but told directly by the teacher) and many housekeeping tasks such as snack preparation, sweeping, dusting, and the organization of materials.

The Elementary School Curriculum

A sequence of themes underlies the Waldorf elementary school curriculum: First grade: fairy tales; second grade: fables and legends of the saints; third grade: Old Testament stories; fourth grade: Norse myths; fifth grade: Indian, Persian, Egyptian-Chaldean, and Greek history and mythology; sixth grade: Roman and Medieval history; seventh grade: Renaissance and the age of discovery; eighth grade: modern history.

According to Waldorf educators, these themes can help teachers address the specific interests and needs of children in their different developmental stages. Let me give an example. They believe that in their first years of life children live more in the spiritual world than on the earth, and that as they grow older, they progressively become more grounded in their earthly life. They also believe that when children are around nine years old, they finally "land" on the earth finalizing their "incarnation" process. This landing on the earth, or as they put it, this "fall from paradise," according to them, is symbolically equivalent to what happened to Adam and Eve when they were expelled from the Eden and suddenly had to start fending for themselves: Children who used to take things for granted suddenly start thinking more about practical aspects of life such as food, housing, and clothing. Based on this understanding, the Waldorf third grade curriculum revolves around epic biblical stories which address both the spiritual world in which children live—and that at some point in third grade will "fall from"—and practical matters of survival. The curriculum includes a unit on farming and building so that children can learn some survival skills and develop some appreciation for all the work that is necessary for keeping us fed and sheltered.

Waldorf teachers usually feel that the sequence of themes that underlie the curriculum of each grade addresses children's interests and developmental needs and stimulates in them the right things at the right time. See for instance the following quotes by different teachers:

> Every year I enter a grade heading into new material with the children in a new age and wonder: Is this going to really interest them? And every year I watch the students drink it in with their whole being.... I feel that at each grade level we were clearly in the realm that is of true interest that really touches their deepest being. (Samantha, 7th grade teacher).

> We feel that it is right on. I mean, to me, having gone around with a group from first to eighth grade, it is real to me. Each year the curriculum really spoke to the children where they were at developmentally.... Like for instance, when you get into 7th grade and you have the age of discovery. That's what they are experiencing at that time. They are little Galileos, and they're renaissance people ... they discover the world ... and then go out as explorers, discover the new world.... And so in that way it is interesting to study that at that time, because it is something that developmentally you are connected to. All the revolutions are taught in eighth grade because at the age of fourteen you really are in some ways feeling ready to take responsibility for yourself and to throw off the old. There is a connection there and so they enjoy studying at that particular time, whereas if you give it to them at another time it is more abstract. (Lauren , 2nd grade teacher)

Besides the regular academic program which is offered in most schools, the Waldorf curriculum includes a variety of other kinds of classes. At the SWS, which has a typical Waldorf curriculum, starting in first grade students learn to play the recorder and have handwork, games, eurythmy, and two foreign language classes (German and Spanish). Starting in third grade they have singing classes. In fourth grade, they start studying a string instrument. In fifth grade they can either continue studying the string instrument or start learning a wind instrument. They are also introduced to chorus and woodwork classes. In sixth grade they start practicing for the school orchestra. Each group of children from first to eighth grade continues to experience each of these classes in increasing levels of difficulty from year to year. As some of the Waldorf classes have unique characteristics—especially the eurythmy, physical education, and woodwork and handwork—I will explain a few of them in the following paragraphs.

EURYTHMY

"Most simply stated, eurythmy is music and speech expressed in bodily movement. It is also called 'visible speech' or 'visible song,' or 'an art form which translates music and speech into movement.' " (Holland, 1981, p. 13). Eurythmic movements are slow flowing and spiral and generally accompany music or recitation representing different ideas, letters, sounds, musical intervals, and colors, among other qualities.

Steiner introduced eurythmy in the Waldorf curriculum for three main reasons: He sees it as a means of artistic expression and he believes that it can foster the development of and connection between body and soul, and that it can help us harmonize with the forces of the universe (Steiner, 1998). Eurythmy, as the SWS eurythmy teacher explained, is not only outer movement but inner movement. It is the movement of the "etheric body" or, we could say, the movement of our life forces. It works with the feeling life of children and helps them develop coordination, strength, rhythm, and spatial and social awareness.

The eurythmy curriculum in the Waldorf school, like the curriculum of other special classes, is connected to classroom activities. For example, as first graders learn the alphabet letters, they also learn to express these letters with their whole body in eurythmy classes. Or as fifth graders study ancient Egypt, they express Egyptian poetry in two-dimensional planes—as it is represented in Egyptian art—during eurythmy classes.

One of my observations during my research was that the Waldorf students seemed to be significantly more gracious and articulate while expressing themselves in public than students from other schools. I could observe this very clearly in school plays and performances. It is likely that much of this graciousness and easiness to express themselves is related to the eurythmic practices they start early in their education.

Handwork and Woodwork

Waldorf students have handwork classes starting in first grade and woodwork classes starting in fifth grade. They learn to knit, crochet, hand sew and machine sew. They also learn to make little objects such as bowls, spoons, and bird houses of wood. Waldorf teachers believe that handwork and woodwork classes are important not only because they teach children important skills, but also because they educate them to carry through with tasks. In other words, they foster the development of children's will, which is the kind of inner power we need to control our thinking and do what we need to do in order to accomplish what we want to and need to accomplish. Jenny, the handwork teacher at the SWS explains:

> If you are knitting a sweater ... it is a huge project. You might be excited for the first part of the sweater and then in the second half, the only thing that keeps you going is that you know that you need to finish it because you want to wear it. It's basically your will telling you. In the first grade they see this recorder bag and they are so excited to knit it. But after the third row they realize that it is boring and they need to keep going ... and to keep adding to it. You cannot skip corners ... and so that's how they are training their will.

During my observations, I noticed one more important quality of handwork and woodwork classes: Some students who had difficulties in other kinds of classes but had some manual dexterity had a great opportunity to "shine" and feel good about their work. I also noticed that some of the most difficult students in terms of behavior seemed to conduct themselves quite well during these classes. Considering that doing some kind of manual work seemed to help them behave better, perhaps it would be a good strategy to have them do this kind of work during other classes (during the times they need to sit and listen to the teacher or to their classmates). This strategy is actually being used by the Bellwether School teachers, and does lead to good results most of the time.

Physical Education

Physical education in Waldorf Education aims at helping students develop inner strength, coordination, flexibility, and spatial awareness. They also aim at attending to the social and artistic needs of the child. At the SWS children have games in the early grades, and starting in fourth grade, they have gym classes. An important aspect of these classes is a lack of emphasis on competition. Even when they introduce children to sports, the emphasis is more on teamwork and good sportsmanship than on competition.

The games and gym classes are connected to the classroom curriculum as much as possible. For example, while second graders do fables in the classroom, they play games that use imagery from the animal world in games classes. While fifth graders study ancient Greece, they practice running, long jump, wrestling, javelin, and discus in gym classes, which were typical sports in ancient Greek culture. They also organize and participate in a reenactment of the ancient Greek Olympic games, where they not only do these sports, but also learn about some of the principles that governed the Olympic games at that time.

The Breathing In and Breathing Out of the Curriculum

Waldorf teachers believe that rhythm affects the whole being of the child, and therefore, they try to provide for a rhythmic pattern in the curriculum balancing what they see as breathing in and breathing out activities. In the context of nursery and kindergarten classrooms, breathing in means quieter activities and breathing out means more energetic activities. In the context of elementary school, breathing in are "activities in which students work more with their minds than with their hands," as in math, German, and Spanish. Breathing out activities are those in which "they work mostly with their hands and bodies." A teacher explains:

> We are trying to work with the balance between expansion and contraction, and we are trying to work with that in every lesson, the whole day, the whole week.... It all relates to teaching the child to learn to breathe and find balance.... They come in and they can play the re-

> corder and move and find an out-breath. And then they are asked to listen to draw in and to focus and to move right into their fingertips and write ... and then off they go to the gym for an out-breath. Then they come back and move into the focus of mathematics. Ideally they find that rhythm in their day.... There is expansion and contraction and in the ideal situation this happens all day long. Creating schedules are a big puzzle, not every moment of every day is ideal ... but that intent lives behind the scheduling for the students. So there is a discipline in that, a discipline of being able to go out and expand, and then the discipline to be able to center oneself again and focus and find that balance. (Samantha, 7th grade teacher)

The Waldorf curriculum has a few unique features that are connected to this idea of "breathing in and breathing out." Waldorf educators believe that children are usually more alert, receptive, and able to concentrate in the morning. Therefore, they set aside the first two hours of the morning for the teaching of the main subject of the day. They call this block of time "main lesson." The main lesson starts with a "morning circle" which usually includes some physical exercises, recitations, and music. The morning circle lasts longer and usually includes more physical activities in the earlier grades than in the older grades. These activities, which are of an out-breath type, according to Waldorf teachers, help students complete their transition from home to school and to focus on the next section of the lesson that is considered the most academic part of the day (the in-breath part of the main lesson). After the main lessons they have snacks and recess for an "out-breath" and release of physical energy.

Another feature of the curriculum connected to the idea of breathing in and breathing out is that Waldorf educators let their students "go to sleep" with some of their school experiences. In other words, they usually let the children go home with their new experiences before asking them to intellectually analyze them. They believe that while children play, rest, and are at sleep, the things they are exposed to are processed and transformed within their being.

Another feature of the Waldorf curriculum related to the idea of breathing in and breathing out is "block teaching." A same subject is pursued in a block of several weeks (usually three to six weeks). The blocks follow each other in a sequence that ideally provides for contrast and variety. Later in the year and/or in the next years they come back to the same blocks, which are taught in increasing levels of difficulty. The disciplines that require a constant practice such as English and math are taught in blocks like the other disciplines, but are also supplemented on a regular basis outside main lessons. According to Waldorf teachers two important benefits of the block method are that (1) it allows them to explore a same subject in depth for an extended period of time; and (2) the interval between blocks allows what is learned to be fully processed.

One could question the block method based on the assumption that without constant repetition, children are likely to forget what they learn. But Waldorf teachers believe that "the process of 'forgetting' is part of the process of assimilating the material, of sorting out the 'facts' from deeper knowledge" (Holland, 1981, p. 6). Therefore, they explain, while children might in fact forget facts and knowledge in the intervals between blocks, the very essence of what they learn sinks into their whole being and becomes part of them. Also, according to them, while students "forget" the subject matter for a period of time, when they get back to it, they are more able to make fresh connections and combinations between the different things they had been learning.

The rhythm of the curriculum and the breathing in and breathing out of its daily life is marked not only by time, but also by ritualized procedures between activities such as singing with movement, the reciting of verses, and/or the lighting of candles. The rhythm of year around activities is marked by special events that celebrate commemorative dates and the different seasons of the year (Leichter, 1981).

Fostering the Feeling and Imaginative Life of Children

The focus of Waldorf curriculum in the elementary years is on the feeling and imaginative life of children, and not on their intellectual development. This latter aspect of development is only emphasized in the high school years (during children's third stage of development). In Waldorf education, teachers are seen as artists who should enliven subject matter, materials, and experiences in order to nurture children's feelings and imagination. For them, imagination is the very basis of creativity and true freedom: We need to be imaginative in order to create new ways of thinking and envision different possibilities of action. We also need to be imaginative for accessing "higher order thinking," a kind of reasoning that reaches beyond the sense-perceptible world.

Waldorf teachers do not use textbooks. They try to bring information to life through an oral curriculum that focuses on story telling and imagery:

> We try to give children pictures that have a certain life and children can live into and can come to them with a certain amount of enthusiasm, whereas if you just give the concept, it's dead. So when you are teaching the letters for instance, you could say this is an "A" and this is a "B," and "B" says "B." But if you develop it out of the pictures ... you start with the story about the bear, and then you draw the bear, and then you can see the form of the "B" and the bear, and you can sound the word and hear the first sound. The child has gone through a very enlivening process ... much more interesting than saying this is a letter "B." (Lauren , 2nd grade teacher)

According to Waldorf educators, as children listen to stories, they create pictures in their heads and hearts which live in them for a long time. Besides, they claim, these pictures not only assist them in understanding concepts and moral principles, but also help them get in touch with deeper levels of reality that can be accessed only through imagination and thinking.

In Waldorf education, in the early grades, the teachers will avoid scientific explanations to facts of life. For instance, if children ask their teacher "Why does it rain?" the latter will not give them a direct explanation about the water cycle. Instead, he/she is likely to tell them a story that talks about this cycle in figurative ways that can potentially open them up to different interpretations. One teacher commented,

> You might for instance tell a story about a little drop of water that fell down as rain into the stream at the top of the mountain, and how this drop of water joined other drops of water by tumbling and splashing and spilling down the mountain, and how it rushed ever faster and faster, and then it slowed down as it came into the river and it saw

> many things along the banks, and it eventually came into the ocean and then it was so hot that eventually the sun fairies came and pulled it up. Then you have a picture of the water cycle, a very imaginative picture. (Lauren, 2nd grade teacher)

Waldorf teachers—especially in the early grades—tell their students a lot of stories that involve fairies, gnomes, angels, and saints. Underlying this approach there is a strong belief in a deeper reality which lies beyond anything our senses can attune to. These stories, they believe, potentially help students keep themselves open to this reality through their imaginative and feeling life.

Waldorf educators make use of a lot of biographies and dramatic plays in the curriculum. These activities, they claim, promote not only knowledge, but also can foster students' feelings and imagination. As children study biographies and participate in dramatic plays, they learn to see knowledge and information in context. This potentially helps them better understand the issues at stake in difficult situations as well as help them to empathize with different people. For example, a seventh grade teacher who I interviewed told me that while she was developing a unit on Galileo, she asked her students to role-play his trial. According to her, as students put themselves in different shoes in the dramatic play, they better understood the complexities involved in the different perspectives in the trial, and felt some empathy for the different people who participated in it.

The Waldorf approach to children's imaginative and feeling life can be quite effective in promoting a degree of emotional involvement between students and their experiences in school, which possibly helps them focus and learn. This approach is also possibly important in helping prevent violence. As J. Miller (2000) remarks, there is evidence from research that children who are more imaginative tend to be less aggressive than children who are not. The ability to be imaginative can help them empathize with other people as well as to figure out different possibilities and alternatives as routes for thinking and action.

The Waldorf focus on the imaginative and feeling life of the child challenges two basic mainstream assumptions: (1) that we should teach children to separate emotions from thinking; and (2) that we should teach them the scientific "truths" about things as soon as possible in their lives. According to the Waldorf framework, while these assumptions are promoting some important scientific and technological advances, they have also been promoting a dead worldview which isn't able to fulfill our innermost needs for meaning and connections.

Promoting the Arts and a Sense of Beauty

The Waldorf curriculum fosters in students a sense of beauty and teaches them to express themselves in healthy and positive ways through various artistic means. The physical environment provided in Waldorf schools is usually pretty. The classrooms are decorated with numerous plants, and depending on the season of the year, also with flowers. The windows are usually big and provide for an abundance of natural light. Most, if not all materials, from manipulatives to desks, are made of organic elements (usually there is no plastic in the classrooms), which along with the plants, give the classrooms a soft and calm atmosphere and sense of closeness to nature. Everything in the classroom has a special place, and objects are kept in order, giving children a sense of organization and neatness. Children's art work is carefully displayed on walls and shelves, decorating and embellishing the rooms without overcrowding them.

The Waldorf curriculum promotes the aesthetic and artistic development of children both through the beauty of the school's physical environment as well as through special classes such as woodwork, handwork, music, and eurythmy. It also promotes this kind of development through activities such as drawing, painting, dramatizations and modeling which are quite often present throughout the school day. For example, often when the teachers present new concepts and/or tell stories to children, they ask them to draw, paint, or dramatize some aspects of the concept/story they had worked on before doing any kind of academic work with it.

The sections in which children draw, paint, or do any other kind of artistic activity are usually treated with the same seriousness as the sections in which students do more academic types of work such as reading or writing. In the Waldorf context, quality is more important than quantity, and children are usually given a considerable amount of time to do their work. They are expected to do their tasks carefully, almost as if everything they did were sacred.

Waldorf educators believe, according to the teachers I interviewed, that among other things, artistic experiences are a means to refine and express feelings. Furthermore, they say, these experiences "engage children's imagination," "nurture creativity," "develop children's eyes and ears for beauty," and "foster eye-hand coordination." Also, and very importantly, they enrich and give depth to experiences and help students assimilate and retain them in their memory. Trostli (1998) explains these last points:

> In Waldorf schools the arts are not taught for their own sake, but because they allow a child to experience a subject on a far deeper, richer level than the intellectual one.... Artistic experiences usually leave lasting impressions in a student's life. Information can be gathered or retrieved, but the experience of the subjects through individual work and through the arts builds a foundation in the soul that will enrich all further learning and the whole of a student's life. (p. xxviii)

The arts, as Trostli (1998) explains, foster a connection between subject and object and integrate cognitive, emotional, spiritual, and kinesthetic learning. Besides, as one of the teachers I interviewed remarked, they are also an excellent means to promote children's socialization. See for instance the following quote in which a teacher talks about how the plays she and her class put together every consecutive year promoted her students' social development:

> I experience them [the plays] to be the greatest vehicle for transformation for individuals and for the class on its journey. For each to take on a role and take a part and support one another is a remarkable experience year after year as a group. And by now they became the directors, the stage managers, the lighting crew, created their own sets, determined their own costumes. It was incredible to see them function as a group.... I literally know that if they had to leave the school ... each could function as a really contributing member of a group with the confidence that "we can pull this off, we can do this!" (Samantha, 7th grade teacher)

Poetry is a fundamental part of the Waldorf curriculum and is present in the daily life of every classroom. Amy, the 8th grade teacher explains in a few words the Waldorf focus on poetry:

> The rhythm is connected with breathing and strengthens our breathing mechanism. The imagery is rich and fosters imagination. Reciting poetry also fosters proper speech, develops vocabulary.... You are actually exercising the brain, you are developing your imagination, you are developing your ability to sequence, you are working with speech which is really an art, so it is incredibly stimulating. And then of course, you are working with very fine material, you have it living in you wonderful imagination.

The richness of the artistic life of a Waldorf School can be verified whenever a visitor walks into it. During *all* my visits to the SWS I would hear and/or observe groups of students playing some kind of instrument and/or see children coming and going from rehearsals for plays or other kinds of performances. At the SWS, each classroom puts together one play to be presented to the school community every year. The quality of their plays, as I have personally observed, is astonishing compared with what we see in other places with children of the same age. During all the performances I have been to, they wore beautiful hand-crafted costumes, played their own music, skillfully danced and sang (in solos and/or in rounds), and used well-developed stage and lighting techniques.

Fostering Movement and Embodied Learning

> We reject utterly that most pernicious modern way of thinking, where the things of soul-spirit are perpetually explained without any references to the body, and then on the other hand—as a perfectly natural corollary—all that has to do with the body is spoken of in grossly materialistic terms. In reality, neither the one nor the other exists as such; what we have in life is the confluence of the two. (Steiner cited in Reinsmith, 1989).

The development of the body and mind go hand in hand in the Waldorf curriculum. The body is present not only in games, gym, eurythmy, handwork, and woodwork classes, but also in the regular classroom activities. In the early grades, for instance, the teachers and students actually take about 20 to 30 minutes in the beginning of the main lesson to do various kinds of activities that involve physical action (e.g., singing along with movement, walking on a rope, jumping, clapping, stumping feet). After that they also usually do a times table game which is accompanied by questions, answers, and rhythmic movements. For example, the teacher will call "7?" and the children will answer "7 x 1." The teacher will then say "14?" and the children will answer "7 x 2." Next he/she will call "21?" and the children answer "7 x 3," and so on. This sequence of questioning and answering is accompanied by a complex sequence of energetic clapping and feet stamping. In between different activities and also at the end of almost every class during the day the teachers and students sing songs accompanied with some kind of movement which apparently help them transition to new activities/classes.

Even the most academic classes in the Waldorf curriculum tend to include some form of physical movement, either through role-plays, games, artistic work, or little outside trips to observe natural phenomena. For example, if teachers and students are working on a unit about time, they are likely to go out to observe the position of the sun in the sky in different periods of the day, to draw different images that deal with the different positions of the sun and the various colors of the sky in the different times of the day, and to build their own clocks on cardboard. Only after numerous "embodied" experiences with the concept of time will they learn to read time as it is expressed in real clocks.

Movement can add a different dimension to learning, and this is something most of us cannot fully realize until we experience it. I personally had two experiences at the SWS that strongly spoke to that. The first of these experiences occurred when I was observing the second grade classroom and finally decided to join in with the group in the times table exercise I described two paragraphs above. Only when I participated in the exercise did I realize how different it felt to do math with my whole body instead of doing it only mentally. Only then did I realize how challenging and fun it was to be a part of that exercise and how helpful that could be to support people in learning and memorizing the times table. As a teacher explained,

> When you have experienced something through movement, you remember in your body that you had it. I think that's tacit. Your body knows it.... When you learn the times table, when we clap them, step them ... it is alive. Then if you forget something you clap and your body remembers.... (Helen, 3rd grade teacher)

Another powerful experience I had with movement relates to an observation in the 5th grade classroom. This class had been studying Ancient Egypt and the teacher asked each student (including myself) to model a replica of the Sphinx. While the teacher guided us step by step on how to accomplish the task and while we worked on our replicas, I felt really connected with my own memories about Ancient Egypt and with the children around me who quietly shared insights on how to do the work. For some reason I cannot explain, the experience of modeling the Sphinx with my own hands—and the connections I made while modeling it—got me interested in Ancient Egypt in a way that I had never been before. Now whenever I see anything related to it, it instantly draws my attention and I want to know more about it.

Once during a conversation with a Waldorf fifth grader, I asked her how she felt in relationship to her peers in other schools. She said she felt that her friends in other schools apparently were exposed to more knowledge and information than she was. But she also felt "they didn't know very many things as thoroughly as she did." The comment of this student seemed to be right on to me. According to my observations, it seems that while the Waldorf curriculum doesn't cover the same volume of information as some good traditional schools do, it does promote a depth of experience that is *not* common to other kinds of schools. A depth of experience that comes from the stories, dramatizations, games and artistic projects which are intertwined with the topics being studied.

There is one problem with the Waldorf approach that is related to this discussion though. While it is true that the Waldorf curriculum promotes a certain depth of learning through embodied experiences, it prevents a different sort of depth: The kind of depth that can only be achieved through inquiry and research. As I will further explain in a different chapter, Waldorf teachers believe that children in the early grades are not ready to and should not be asked to weigh different viewpoints and make judgement calls on their own. Therefore, they usually do not give primary school students many opportunities to raise and investigate their own questions. Consequently, the children end up relying very much on the teacher as the main, when not the only means of information, which can eventually confine them to some of the limitations and biases of their teacher. This situation may in fact be aggravated considering that in Waldorf education, as I will explain next, the same teacher ideally stays with the same group of students during all the elementary years.

Eight Years in a Row with the Same Teacher

> During the first seven years children establish the foundation of their relationship to the world around them. From the age of six or seven to fourteen, they undergo a tremendous process of inner growth and begin to discover the world within themselves. Whereas the young child learns primarily from his environment and by imitating the words, gestures, and actions of those around him, the child of elementary school age learns through his feelings for those who teach him. (Trostli, 1991, p. 342)

According to Waldorf educators, children start school with a huge eagerness to learn and with a tremendous faith in the teacher. Also according to them, children need close, intimate, stable, and continuous relationships with a mentor for a proper development and socialization. After all, they say, children's learning—especially in elementary school—is based, to a great extent, on the kinds of feelings they have for the teacher. Based on these beliefs, in Waldorf education, the class teacher ideally stays with the same group of students from first to eighth grade. Obviously this is not always possible since problems might emerge throughout the years (e.g., an illness might force the teacher take a year off work or he/she might have to move away).

Some of the advantages of the eight-years-in-a-row approach—and the fact that the teacher stays with the same group of students for a long period of time—are:

- Since the teacher stays with the same class throughout the years, he/she can assure consistency and continuity in the school experiences.
- People can have a sense of stability and continuousness in their relationships.[1]
- Since people (parents, teachers, students) spend so much time together, they have numerous opportunities to develop a strong sense of community.
- Members of the community are compelled to learn to work together regardless of their problems and differences (it is easier to get away from this kind of learning when we know we will be in a different group every year).

From my perspective, the eight-year-in-a-row approach has two main weaknesses:

- Sometimes the teacher and a given student (and/or the latter's parents) are not able to work out their differences and solve the kinds of problems they have with each other. As a result, they either have to forcefully put up with each other or the student might have to leave the school (most Waldorf schools have only one classroom for each grade, which means that students usually do not have the option to change to a different classroom).
- It may be hard for some teachers to master all the knowledge and skills they are supposed to cover from first to eighth grade. While students may indeed learn important lessons from watching their teachers struggle to master new knowledge and skills every year[2], not always are teachers able to grow as much as is needed to be truly effective in their teaching. Some Waldorf schools are starting to recognize this as an issue and are trying to address the problem by having specialist teachers working with children in specific academic disciplines. The SWS for example, has hired a math teacher to work with 6th, 7th and 8th graders. The class teacher is still responsible for introducing students to mathematical concepts during the main lessons. But the math teacher works with the children on specific skills during other blocks of time.

Waldorf advocates believe that the benefits of the eight-year-in-a-row approach outweigh its problems. See for instance the following passage by Edmonds (1992):

> "How long does it take an experienced teacher to know thirty children or more—really to know them, understand them and be able to enter into their intimate needs?" And the contrary, "How long does it take children to grow so accustomed to the quality of mind, the temperament, the mannerisms of a teacher that they feel happily anchored, understood, and secure?" And what can it mean in a child's life to have to make a new adjustment to a personality every year again? What can be the effect on a young child of being thus uprooted and transplanted year by year during the most formative years of its life—a different quality of discipline as well as all else? ... And what does it mean to a teacher to be forever dealing with the same age group, the six year olds, the nine year olds, and so on, knowing hardly at all what went before and without responsibility for what comes after? Where is the sanity in this which is accepted as the universal practice? We have seen that a child's life is not only a succession of years—it is a life-developing process, and like all such processes there are nodal points and intervals and crises, that is, times of vital transition which have to be specially known, prepared for, met and carried over. All this is part of a teacher's profoundest service, and how can it be achieved without the kind of continuity with the children in their growing, and with their parents? How, in these turbulent times, can we hope to find people with inner strength of security, feeling strongly anchored in themselves and in their tasks in life, if they have not experienced that anchorage and security in their growing years? (p. 54-55)

While considering the points above, I still do not have a clear position about the eight-years-in-a-row approach. It seems to me that perhaps a good alternative would be to have the same teacher working with the same group of students for three or four years rather than for eight years. This would enable us to keep the positive aspects of the approach and at the same time, minimize the above mentioned problems.

The Curriculum Beyond the Classroom

Waldorf education deals with the child as a whole being. Therefore, Waldorf schools are concerned not only with what happens with the child in the classroom, but also outside it. For example, they ask parents to maintain a consistent rhythm in their children's lives and to protect them from overexposure to popular culture so that children can have time, space, and energy to digest their school experiences. At the SWS, they ask parents not to expose children to TV and other electronic media until at least 4th grade. When parents feel this is not possible, they are asked to avoid these media at least during school days so as not to interfere with the work they are trying to do at school. A teacher explains:

> We try to give children something that they can take into their sleep life, something of substance.... If they go home and watch television that night and draw all kinds of images from the TV, that will really work against what we've been trying to give them during the day. (Lauren, 2nd grade teacher)

Waldorf educators see TV and other electronic media as potentially harmful for children for several reasons: They do not stimulate an authentic and creative imaginative life, offer virtual rather than real life experiences, and almost always aim at selling a materialistic and consumeristic worldview and lifestyle. This all goes directly against the kinds of values promoted by Waldorf Schools.

According to many of the teachers I interviewed, they do not want, by any means, to control their students' private lives. However, they do feel they have to intervene when they sense that what is happening at home is interfering with children's learning and behavior in school. When this is the case, they might for instance, invite parents for a meeting or a series of meetings where they can discuss both the family's goals and the school's goals for the child. When these goals are in agreement, they (teachers and parents) often try to work as a team to achieve them. When the goals are not compatible, the school is likely to ask the parents to look for another school for their child, one that is more in agreement with the family's educational philosophy.

In order to educate the parents about their educational philosophy and practice, the SWS publishes periodicals and offers workshops for parents and for the school community on a regular basis. About once a month the teacher and the parents of each class meet to discuss educational and organizational issues. During these meetings the teacher usually leads a discussion about his/her work with the students and/or directs an activity that he/she has recently developed in the classroom so that the parents can have a "feeling" for what their children have been doing in school. During these meetings, they also discuss organizational issues such as how to raise more funds for the class or how they can get organized for the next field trip. Besides, they also talk about things of common interest and try to support each other in the difficulties they are having with their children. For example, during a parent/teacher meeting of the second grade group which I had the opportunity to observe, the "class parent" (a mother who had been elected by the parents to represent them in the school and to serve as a liaison between them and the teacher) raised the following question to be discussed during part of the meeting: "What are the most difficult challenges you are facing with your 2nd grader at home?" The parents talked about this subject and exchanged experiences and strategies to support each other in dealing with the problems they had been facing with their child at home. The teacher also intervened with some insights which contributed to the discussion.

Waldorf parents often do a considerable amount of volunteer work in the school. Some, for example, help in the handwork classes teaching children how to knit, crochet, sew, or embroider. Others volunteer in small reading groups or in the production of plays, organization of social events, school festivals, fundraisers, and field trips. They seem to enjoy doing such kind of work since, as some of them commented with me during interviews, while they are in school, they learn not only about their children and the school's educational philosophy and practices, but also about themselves.

It is interesting to note that many of the parents I interviewed mentioned that the Waldorf School is as much of an educational experience for them as it is for their children. For example, some of them had to educate themselves not to watch television so they could keep their children away from it. In some cases they also had to reconsider their eating habits so their children could adapt to the wholesome food served in school.[3] Some parents also started to study Anthroposophy as they read and learned more about Steiner's work. A few parents also mentioned they learned from the other parents as they helped take care of each other's children. As their children stay together in the same classroom for many years, they end up spending a lot of time together not only in school but also outside school. Along the years the families get to know each other well and to respect their differences. After all, they cannot escape from dealing with and from learning to work on their differences since their children are "stuck" together for a long period of time.

BUILDING COMMUNITY

Various factors contribute to the building of a strong sense of community at the SWS. Many I have already discussed, as for example, the eight-years-in-a-row approach, the weekly faculty meetings, the regular meetings between teachers and parents, the school-wide workshops, and all the volunteer work done by the parents. There are other factors as well.

Children sing, play musical instruments, and present at least one play every year together. While singing in rounds, playing instruments together, or interpreting their role in plays, they need to sing/play/interpret their part and still pay attention to the part of others so that they can be in tune with the rest of the group. If one element doesn't do his/her part well or sticks out, the work of the whole group is affected. This strengths their understanding that, if they are to reach common goals together, they need to listen to each other, respect one another, and follow certain rules.

Another factor that contributes to the building of a sense of community in the school are the weekly "morning assemblies." All teachers and students of the elementary school meet every Monday early in the morning. They usually start these meetings by singing a song and/or saying a verse together. Then a group of students makes some kind of artistic presentation (e.g., they might play a musical instrument, recite a poem or sing a song). And lastly, each teacher tells the whole school what their class will be doing during the week.

> The morning assemblies bring the whole school together, a view of the whole of which they are all a part of. It a sort of reminds them what the school is about. It's good to know what everyone is doing. The older students remember what they used to do when they were younger and the younger look forward to doing when they grow older. (Lauren, 2nd grade teacher)

As students say verses and sing songs together, watch each other's artistic work, share what is going on in their classrooms, and remind themselves of their common past or look forward to their common future in the school, they develop a sense of being a part of a larger whole. Part of a community which they are learning to know, to work with, and to respect day after day while they are in school.

Other factors that promote a sense of community in the school are the various social events that the school and parents put together such as school plays, concerts, seasonal festivals, and fairs. There are also the faculty meetings where the teachers discuss issues of mutual interest, work on their personal and professional development, and support each other on the problems they are currently facing in their classrooms.

The Waldorf curriculum addresses the whole child. Everything children learn speaks not only to their head but also to their hands and heart. For example, if they are to be introduced to letters, they will physically do something with these letters (e.g., they will make the letters with their arms on the air, they will walk over the letters on the floor, they will play games with the letters, and they will do eurythmic movements representing them). They will also draw the letters with crayons and model them in beeswax. Besides, they will dramatize stories, look at images, sing songs, and recite poems that relate to the letters. All these activities will help them relate to letters on a physical and emotional level. Only after they establish a strong relationship with the letters both in the physical and emotional level, will they be asked to intellectually deal with them as symbols to represent ideas in reading and writing.

The Waldorf curriculum is integrated in different ways: There is no clear cut distinction between the different subject matters, and the specialty teachers (handwork, woodwork, games, physical education, music and eurythmy) try to connect their curricula to classroom themes as much as possible. The arts permeate the whole curriculum, and the practice of artistic activities are considered as important as any academic subject.

I still haven't talked about one essential aspect of the Waldorf curriculum, that is spirituality. Considering the importance of this aspect, I will dedicate the whole next chapter to discussing it.

Notes

1. As Trostli (1998) explains, "In an age where relationships so easily dissolve for lack of commitment, a child is strengthened by the knowledge that his class teacher loves him, will stand by him, and will accompany him through this part of his journey through life. Such knowledge gives children faith in the power of human relationships to endure." (p. xxiii)
2. They can learn for instance that through effort and dedication a lot can be achieved.
3. The school doesn't allow for candies or pop sodas in the building. As an alternative, it sells a wholesome vegetarian lunch prepared on a daily basis in the school.

Chapter 9

Promoting a Sense of Sacredness, Gratitude and Reverence

Spirituality in the Waldorf Curriculum

The concept of reincarnation and the idea that we are spiritual beings in search of self-development through a series of bodily existences underlies the whole Waldorf philosophy. Therefore, spirituality is a very central theme in Waldorf education:

> [Waldorf teachers have] this picture that ... there is something that exists beyond just what we are in this moment, beyond our physical body. There is a piece of each one of us that is eternal, and it goes on. You might think of this body of ours as a cloak that we put on.... This cloak is of course fashioned in the most beautiful way out of the substances of the earth ... the substances of the world meaning the chemicals, the air we breath, mineral substances ... and when we are born we take it on, we put this cloak on, when we die we take it off, and yet we

> are still there, we still exist.... After death then there is this whole process of reviewing what has happened and this continuing development through successive lives. So there is a sense that there is a part of each one of us that is eternal and it lives on beyond this particular life. And in a way the children who are in front of us, they are all these eternal beings, and some of them maybe are incredibly developed, and have an incredible wisdom, and there are people who perhaps who we have had something to do with once before in other previous lifetimes. So there is a tremendous sense of respect for what we are doing.... (Lauren, 2nd grade teacher)

In order to ground my discussion of the ways in which spirituality is actually expressed in the daily life of the Waldorf curriculum, I will offer a little snapshot of part of the daily routine of the second grade teacher, Lauren Walker, in whose classroom I spent most of my time while developing my research at the SWS.[1]

Lauren would come early every morning to prepare her classroom for the day. At eight o'clock, she would go to the teachers' room where she and some other teachers would read and comment upon a little piece of the "Calendar of the Soul"[2] by Steiner. She and the other teachers would then recite the following verse together (from Steiner's dedication of the first Waldorf School in Stuttgart) before they went back to their classrooms:

> May here rule what Spirit-strength in Love
> May here work what Spirit-Light in Goodness
> From surety of heart
> From steadfastness of soul
> Can render to the young
> For their body's strength in work
> For their soul's intensity
> For their spirit's clarity.
> To this may consecrated be this place.
> May the heart of youth here find
> True human care
> Endowed with strength
> Devoted to the light.

This daily routine helped Lauren connect with her highest purposes as a teacher and to focus on the work she was about to start every morning. At 8:20 a.m., she would stand by the door and make eye contact, shake hands, and say good morning to each individual student as he/she entered the classroom. At 8:30 a.m., she would delicately play a few keys on the xylophone to let children know that it was time to sit down so they could begin their day together. She would then wish the whole group good morning, which was followed a chorus that said "Good morning Mrs. Walker." After this little ritual, Lauren and her students would read the calendar from the board and review the schedule of classes and activities planed for the day. They would then do some stretching and physical exercises and recite the daily morning verse[3]:

The sun with loving light
Makes bright for me the day.
The soul, with spirit power,
Gives strength into my limbs.
And sunlight shining clear
I reverence, O God,
The source of human strength
Which though as gift of grace
Has planted in my soul
That I with all my might may love to work
And ever seek to learn
From thee stream light and strength
To thee rise love and thanks.

After the morning verse they would often sing the following song:

The golden sun
So great and bright
Warms the world with all its might
It makes the darker green and fair
And tends each thing with ceaseless care
It shines on blossoms on stones and tree
On bird and beasts and on your and me.
Oh may each deed throughout the day
May everything we do and say
Be bright and gold and true
Oh golden sun like you.

They would then move their desks and chairs next to the walls and would form a large circle and repeat the following affirmation followed by a sequence of movements a few times: "In me there is light, I have strength in my legs, and love in my heart." After this, they would begin a series of songs, dances, games, and stretching, coordination, and focusing exercises.

The morning circle lasted for about 20 to 30 minutes. Lauren and her students would then put their desks back in rows and would do a series of times table rhymes accompanied by rhythmic movements.[4] They would then proceed to the most academic part of the morning where they worked on specific concepts and/or skills. At the end of the main lesson, the class would do some stretching, a few physical exercises (sometimes they would also play the recorder), and recite the following verse:

> With my work I am through
> Now it shall my soul imbue
> That it may grow and make me free
> Grant wisdom, love, and strength to me.
> That I may be of honest worth
> To all human beings and to the earth

Next there was snack, but before eating they would always sing a song and/or say a verse thanking for the things they had and for the food they were about to eat:

> Oh the lord is good to me
> And so I thank the lord
> For giving me, the things I need
> Like the sun and the rain and the apple seed
> The lord is good to me
> Alleluia!

> Earth who gave us this food,
> Sun who made it ripe and good.
> Dear earth, Dear Sun
> By you we live.
> Our loving thanks to you, we give
> Blessings on our food and on ourselves!

Above we can see only a little piece of the day of one Waldorf teacher and her classroom at the SWS. Nevertheless this piece provides us with some good basic data to discuss some ways in which spirituality is expressed and nurtured in the daily life of a Waldorf School.

Waldorf schools do not subscribe to any specific religious tradition. Nevertheless, as Waldorf Schools select their new teachers, they usually require that they be committed to their own personal—and we could also say, spiritual and professional—development. After all, how can they help others in their personal/spiritual development if they are not working on their own? All the Waldorf teachers I interviewed claimed to have some kind of spiritual practice. The least that some of them did was to try to set apart some time every day to meditate on their daily practice and on issues that were going on in their classroom with specific students (this kind of meditation was actually a recommendation that Steiner made to all his teachers). A good example of how Lauren and some of her colleagues nurtured their own spirituality can be seen in their willingness to meet every day early in the morning to read and discuss a part of the "Calendar of the Soul" together. Those meetings helped them focus on their inner-self and reminded them of their higher purposes as educators.

One of the ways in which Waldorf schools nurture spirituality is through the daily eye-to-eye handshaking with individual students as they enter the classroom. As teachers look right into the eyes of their students, shake hands with, and wish good morning to each child, they let students know that they acknowledge, value, and respect each one of them as an individual. This small ritual potentially helps children realize that each one of them is acknowledged as a unique person as well as help them transition from the outside world to the sacred space of the classroom which they share with their teacher and peers.

Another way in which the Waldorf curriculum nurtures spirituality is through the extent to which the teachers create opportunities for children to develop a sense of wonder and reverence for the world around them and for a higher power that exists beyond each one of us. As we could see in the example above, they were constantly thanking for the good things in their life in their verses and songs. They thanked and reverenced for example, the food they were to eat; the sun, the earth, the air, and the water for making life possible; and the beauty, love, and friendship for making life being worthwhile being lived.

A common practice of Waldorf teachers who work with younger students is to tell them stories about materials they are about to use and activities they are about to engage in. For example, if a nursery or kindergarten teacher is to make a wool doll with students, he/she is likely to tell them a little story or sing a song that talks about where the wool came from. Usually, after such stories and songs, the child no longer sees the wool just as a thing. It becomes a special gift from the sheep, a gift that she should be thankful for and should treat with care and respect. The Waldorf community also organizes numerous festivals where the school staff, children and families celebrate nature. They celebrate for example, the different seasons of the year. After such festivals, for those who participate in them, winter, spring, summer and fall are no longer just seasons, but presents from heaven that enrich our lives with different kinds of gifts such as snow, flowers, the warmth of the sun, and colorful leaves.

A sense of awe, gratitude, and reverence, Steiner claims, is extremely important since this can help us "see into things." As we learn to deeply observe and appreciate a flower, for instance, we are more likely to be sensitive and attentive to the forces that keep it alive and make it beautiful to our eyes. As Steiner contends, "reverence awakens in the soul a sympathetic power through which we attract qualities in the beings around us, which would otherwise remain concealed" (Steiner, 1976, p. 14).

Spirituality is also nurtured in the Waldorf curriculum as it cultivates mindfulness. This quality is cultivated, for instance, as the teachers provide students with long blocks of time to complete their assignments and focus on the quality rather than on the quantity of their work. The teachers also constantly try to bring students' attention to the moment through physical movement, music and poems immediately preceding and following specific activities.

Waldorf schools potentially foster children's spiritual life also as they protect them from popular culture, mainstream values, and overflows of information. This kind of protection presumably gives children more freedom to be who they are without too much influence from materialistic and consumeristic ideals. Besides, this may help them keep alive a sense of wonder and an ability to appreciate the beauty of small things. This ability, as we know, is something many of our children no longer have, since they have been so constantly bombarded with huge amounts of information and fast moving images that tend not only to distort reality but also to make facts of our real life seem too insignificant, small and/or too slow to be observed and appreciated.

Another way through which the Waldorf curriculum nourishes spirituality is through the telling of stories that bring natural elements to life and involve fairies, gnomes, saints, and biblical passages. These stories implicitly promote in students a "mythical consciousness, one that affirms the living reality of the elements and creatures of the natural world" (Trostli, 1998, p. 98). They offer metaphors that can help keep children open to the existence of deeper levels of reality, levels that are usually not perceptible to our regular senses.

The Waldorf curriculum also cultivates spirituality to the extent that it nurtures in children a sense of wholeness and connectedness with their own self. It does so as it values and addresses all aspects of the child—physical, emotional, intellectual, social, spiritual, and aesthetic—and as it exposes students to a rich curriculum that leads them to explore different potentials and deal with their personal strengths and weaknesses. This potentially promotes self-knowledge, self-awareness, and helps them transcend personal limitations.

The Waldorf curriculum also nurtures in children a sense that they, other people, and everything else in the universe are interconnected and interdependent. This is nurtured, for example, through: (1) poetry and music (e.g., see the words of the songs and poems quoted above); (2) numerous trips to natural settings (e.g., nature walks, canoeing, sailing, and camping expeditions); and (3) the kind of artistic work that is done in school (e.g., the younger children are not allowed to use pencils, only crayons, which necessarily leads them to explore the fluidity and connectedness of the objects in their drawings—the crayons do not allow them to draw clear and definitive boundaries in their drawings).

The science classes in Waldorf Schools also foster a sense of interdependence and interconnectedness. Children in Waldorf Schools study natural phenomena in context. For example, if they are to learn about a plant, they will not pull it apart from its natural habitat. They will look at it in its relationship with other forms of life, with the earth, with the sun, with the moon, with the air, and with other forces that keep it alive. In the Waldorf approach, as Mary Henry (1993) claims,

> People are seen as "part of nature," and the environment is to be protected and valued, not mistreated and exploited as it is in contemporary society.... Waldorf School encourages students to look at plants, the sun, moon, stars, and "beasts that feel and live" as parts of an organism, a cosmos in which they are located....The Waldorf School manner of perceiving the world is not one of control over one's environment, changing it and mastering it through a better technology and science, but one of developing individuals who learn from and are responsive to their environment. (p. 203)

Spirituality is a very central theme, probably the most important theme in Waldorf education. The whole Waldorf curriculum works to foster a sense of connectedness and interdependence with other people, with the natural environment, and with other living beings. The curriculum also promotes a sense that there is a higher power that exists beyond ourselves—a sense that there is more to reality than our regular senses can perceive; a sense of wonder and appreciation for the beauty around us; a sense of gratitude and reverence for our gifts and for all the things that are a part of our lives; a sense of being whole and connected with one's inner being; and a sense of being present in the moment and deeply feeling one's experiences. The Waldorf routines—e.g., morning verse, shaking hands with students, and songs and poetry before, in between, and at the end of classes—and school celebrations and festivals promote the spiritual life of the school and hold the school together as a strong community.

Notes

1. Lauren openly received me in her classroom and made time for various in-depth interviews, which allowed me to fully explore her educational philosophy and practice. It was through her that I gained access to the other teachers' classrooms.
2. The *Calendar of the Soul* is a collection of verses, one written for each week of the year.

3. According to the literature, first to fourth graders in Waldorf Schools all around the world say this same verse (with a few variations given the different translations from the original German) in the beginning of the school day. It was created by Steiner and recited by him in the foundation of the first Waldorf School. From fifth grade to high school they say a different verse which is also apparently universal in all Waldorf Schools.

4. I explained how this times table exercise is done in the previous chapter.

Chapter 10

Conceptions of Freedom and Structure in Waldorf Education

> Our highest endeavor must be to develop free human beings who are able of themselves to impart purpose and direction to their lives. (Rudolf Steiner cited in Holland, 1981, p. 1)

Waldorf educators have a profound respect for the inherent wisdom of human nature—especially for our natural inclination to search for our own divinity—and the baggage that each one of us brings from our previous experiences (from this and from other lives). But while they recognize our wisdom, they also acknowledge that we are still imperfect beings with limitations that may lead us away from our personal path of self-development. They believe that as educators, we should guide children in their experiences so that their energy doesn't get dissipated in vain pursuits and so that we can help them "rise above their limitations." According to them, careful guidance is necessary if we are to help children learn to attune to their "inner-teacher" and follow their path towards self-development and freedom.

Freedom, according to Waldorf educators, is the main goal of education. But freedom for them is not necessarily related to the idea of "choices" and/or "absence of constraints." It is a view of the world and a way of being that is grounded in one's innermost being, and *not* on mainstream values or illusions imposed on us by external influences. In other words, it is "a view of the world based on spiritual activity, on action, on thinking and feeling that arise from the individual human spirit" (Steiner, 1999, p. xiv).[1] It is also a way of being that expresses harmony between will, feeling, and thinking, and an ability to intervene between stimulus and response so that we will not simply respond automatically to situations that are given to us. In other words, it is a way of being that encompasses self-control, which is essential if we are to reach higher and spiritual goals, goals which go beyond those presented to us by mainstream culture.

Steiner and his followers believe that the path of true freedom requires an understanding of children's developmental needs. According to them, during the first stage of development (zero to seven), children learn mostly through imitation and example. Therefore, they see our main role as educators as being good role models for children and providing them with a beautiful and organized environment. They presume that during the second stage (ages seven to fourteen) children "have a deep longing for direction and authority and want a "teacher who can 'keep them in order'" (Harwood, 1992, p. 85). Hence, they see our main role as being to guide students step by step in their experiences. They assume that during the third stage of development (ages 14 to 21) children wish to and need to elaborate and assert their own viewpoints. Therefore, they believe we should give them opportunities to exercise their independence, and see our role as to be learned masters who can help them understand and assimilate specific skills and bodies of knowledge, and support them in their journey towards self-development.

According to the Waldorf framework, children should not be encouraged to engage in rational thinking, and therefore, to make judgment calls, before they can "perceive reality as it really is" and "feel it with their whole being." Steiner (1998) explains:

> What the intellect has to say about any matter should only be said when all the other faculties of the soul have spoken. Before then the intellect only has an intermediary part to play; its task is to comprehend

> what occurs and what is experienced in feeling, to receive it exactly as it is, not letting unripened judgment immediately come in an take over. For this reason, until puberty children should be spared all theories about things; the main consideration is that they should simply meet the experiences of life, receiving them into their souls. Certainly they can be told what various people have thought about this and that, but exercises of judgment, too early, about one view or another must be avoided. Thus, children should receive people's opinions with the feeling power of the soul. Without jumping to a conclusion or taking sides with this or that person, they should be able to listen to all, saying to themselves: "So and so said this, and another said that." (p. 42)

Also according to the Waldorf framework children should not be called to make judgments and decisions before they have enough knowledge, experience, and skills to have true comprehension.

> Nothing is more harmful to children than to awaken independent judgment too early. Human beings are not in a position to judge until they have collected material for judgment and comparison in their inner life. If they form their own conclusions before doing so, their conclusions will lack foundation. Educational mistakes of this kind are the cause of all narrow one-sidedness in life, and all barren creeds based on a few scraps of knowledge, ready on this basis to condemn ideas experienced and proven by humankind often throughout long ages. (Steiner, 1998, pp. 41-42)

Waldorf educators believe that we should guide students, since any choices they might make throughout childhood will affect not only their momentary but also their future experiences. A teacher explains:

> I think it is counter-productive [to let children make decisions]. You would be robbing the child from their experience of feeling safe and protected and comfortable in no responsibility—that the adults have the responsibility for your choice, that the adults know what is right and what is wrong for them—and then they can let go of that and be children and be imaginative and free in their imagination. They need to have structure and authority and direction. Choices rob them of their ability to later on, to really know what they want, because now

> they don't know what they want. If right from the beginning they are given choices, the choices they make shape them.... (Rose, 6th grade teacher)

Waldorf educators want to make sure children will be exposed to specific learning situations to which they might not spontaneously expose themselves either for lack of opportunities, because they were not naturally interested in these kinds of experiences, or because they were afraid of failure. They want to make sure children will move beyond where they are in spite of their resistance, fears, and limitations:

> If you are allowed to follow your own inclinations to the "nth" degree, and you have not experienced anything else, it is this narrowing of your possibilities. In the end you can only do that. But if you had to expand your possibilities, the more doors you open, the more freedom you have.... What you are doing now will determine what opportunities you will have later on.... So, yes, you know, it is hard for you now, and you need to work harder at it now, but what we are trying to do is to keep doors open.... It's trying to help the child to overcome their nature so that they have new possibilities, because otherwise, we just get stuck following our own nature and what we are inclined to do and what we can do well ... and we never really discover these other possibilities. (Lauren, 2nd grade teacher)

Waldorf elementary school teachers direct students step by step throughout almost all their experiences at school. Only as children grow older are they allowed a little more independence, but always under the control of the teacher. For example while 6th, 7th, and 8th graders are allowed to choose some of their research topics, their choices are usually bounded by the options given by the teacher.

The extent to which Waldorf educators structure students' experiences can be well exemplified by how they direct the artistic activities they develop at school. I have observed a few painting sections in the SWS 2nd grade. The teacher would walk students step by step through the whole painting section. She stood in front of the room painting, giving them directions, while the students listened to her and copied from her picture. She would, for instance, tell the children how they should hold the brush and rinse it in the water, what colors they should use, and from what direction to what direction they should move the brush on the paper. According to some students I talked to, only once during the whole school year did she let them freely choose what they wanted to paint in class.

According to some of my interviews with alumni, teachers, and students, the degree to which Waldorf educators allow for self-expression in the arts varies from teacher to teacher. But as a general rule, elementary Waldorf teachers tend to guide children's artistic work to a large extent. This obviously generates many criticisms against Waldorf education since many of us think this might "inhibit" the expression of children's natural creativity. Waldorf educators do not worry about this issue. For them, artistic activities in the earlier grades should be a powerful means for "impression" and not for expression. Jeffrey Kane (personal communication, 2000) explains:

> We often view art as a form of catharsis. We ask kids to let things out. Quite often in Waldorf the objective is the opposite. We tend to ask children to experience yellow, red or green, or the gesture of a tree, or the way the blue sky merges with the red of the sunset, or the sublimity and calm you might experience in playing or listening to Bach. It is not so much letting things out but letting the world or culture make its impression on a growing child.

Waldorf educators see the kinds of artistic activities they develop at school also as "pedagogical exercises" to teach children techniques and skills that in time will give them the tools they need in order to become truly free and creative. I have to admit that when I first started observing the SWS, I was personally shocked by the extent to which the teachers directed students' work. However, in time, I came to realize that since Waldorf students are progressively and systematically led to explore different artistic means, tools, and skills, independently of their personal inclinations, as they grow older, they seem to develop a vast and wonderful repertoire of knowledge and abilities to be creative and to artistically express themselves. While I didn't find any independent scholarly work on this subject, according to my personal observations at the SWS, the quality of Waldorf students' artistic work in the older grades in gen eral can be quite astounding, way beyond the level of the work of students in all other schools I have observed.

The Waldorf curriculum follows the form and sequence I presented in an earlier chapter, and the teachers carefully filter what is addressed in the classroom. The teachers will only open space in their classes to address issues that are not planned to be on their agenda—issues that children say they want to discuss—if they verify that (1) the students are genuinely interested in these issues and are ready for the kind of discussions that might emerge from them; (2) the issues are of the interest of the whole class, and not just of a few students; and (3) important lessons can be learned from the discussions that are likely to come about. They believe that the media, popular culture, and adults in general tend to impose on children issues that do not really "live in their hearts," forcing them to deal with "adult issues" too often. The statement below by Davis William[2], one of the founders of and a teacher at a charter Waldorf School well illustrates this last viewpoint:

> If a first grader comes to school mouthing words that their parents have fed them or their friends or older brothers have fed them, that is not coming from the child. You know, I have seen seven-year-olds coming in and saying "nuclear power is bad" and they have been wearing these t-shirts. They don't know anything about nuclear power and they don't want to talk about nuclear power. They are telling you what their parents have instilled in them as a value.... That is not coming from the heart of the child. So you need to distinguish what is coming from the heart of the child. Is that really the child's concern?... [We need to] keep little children free from the burden of dealing with direct strong tragic or overwhelming experiences in life.

Only as children grow older and develop a more solid basis on which to build good opinions and viewpoints do Waldorf teachers allow for certain topics to be discussed in the classroom. For example, while interviewing Mark, the 4th grade teacher at the SWS, he stated that he normally would not lead a discussion about racism with his fourth grade students unless this was a really boiling issue in the classroom. But he would probably have this kind of discussion with eighth graders, since by eighth grade students would have already studied the history of slavery and imperialism, and would have a good basis to engage in rich conversations about the topic and a good condition to develop independent standpoints.

Waldorf schools tend to protect elementary school children from the harsh reality of the real world until they feel children are prepared to deal with it. Holland (1981) explains:

> In the same way that the parent takes responsibility for sheltering the infant from the physical dangers of its environment, we see a positive role in guiding grade-school children in their confrontations with outside temptations and with the harsher realities of modern society. The child in whom the values and the ideas of a better way of life are instilled at this age is then able when critical faculties awaken in later years to make the judgments necessary for dealing rationally and ethically with a less than perfect world. (p. 12)

Waldorf teachers in fact tend to "protect" children from real-world problems (e.g., poverty, hunger, abuse, or violence) even when some of their students are living these problems in their everyday life. Davis William (personal communication, 2000) explains:

The whole structure of the classroom in the Waldorf School is supposed to support the child.... It's a very protected environment.... It's more like a village, a community that is created in the classroom and that brings stability and a sense of home for the child. Each of the children are on a journey along with that child, and they feel that they are friends, but in a deeper sense than just friends that you play with. These are friends that you play with, that you go to school with ... you sing with.... All of that has a deep impact on the child. They are in plays with each other every year, they're in dramatic productions, and they are experiencing a very deep relationship, hopefully with their teacher, with one another, and with the curriculum. *That provides an environment that the child not only feels as safe but also as nurturing and alive filled with an element that really feeds their soul. So they take refuge in the classroom regardless of what is occurring out in the world, whether it's hunger or abuse.* And if children come to school hungry, the other children will pull from their lunch and feed them. That happens all the time. So *it doesn't cure the ills of society or the ills of their family, that recreates an op-*

portunity to enter into a different world that will always appear in children's memories and live as hope and hopefulness (my emphasis).

As we can see from the quote above, Waldorf educators want children to feel that, even though things might not be going well for them outside school, at least in school they can feel safe and happy. They want them to be able to be children and to appreciate the beauty around them—even if only in school—so they can develop a sense that life is worthwhile being lived.

> We [Waldorf teachers] want to foster in children an appreciation for the beauty around them and a sense that the world is a good place, so that one day they will be able to look back, when they are conflicted with problems in their lives, and remember what it felt like to be a child and that the world is a good place. That the world is a place that is worthy being and working for (Maria, nursery teacher).

Some Waldorf Schools tend to be much more structured than others. The Sunshine elementary Waldorf School is an example of a very structured school. It even has a dress code that determines, for instance, that students should *not* wear clothes with "distracting colors and patterns" or with graphics and writing. It also determines that they should not wear short shorts, halter-tops and tank tops, or clothes inside out. Besides, it forbids them to have long fingernails or to polish them, to dye their hair or use "odd hair-styles," and to use excessive makeup. A teacher explains:

> If it is something that prevents students from doing work that is required, then we establish a policy that it needs to stay at home. For example shirts that have writing or pictures on them. They might not be distracting to the student that is wearing it, but it is certainly distracting to other people in the room and creates visual garbage.... In terms of fingernail polish or long fingernails, that can become the object of what the girl has in mind, and she gets more intrigued with [that] ... than she does with what she is supposed to be doing. In addition these longer fingernails make it difficult to hold a pencil or to do geometry or to hold a paintbrush. These are things that get in the way of what needs to be happening. Short mini-skirts are not only distracting to

> some people, but also inappropriate for gym class.... If they want to wear those things on the weekend, if they want to paint their nails on the weekend, that's up to them. In terms of being here for school, they need to be focused on their work....! (Amy, 8th grade teacher)

The SWS rules for the high school students are not as strict as they are for the elementary school students. Waldorf teachers believe that high schoolers need to have some leeway:

> We don't really approve of dyed hair in elementary school. The high school doesn't make such a big deal out of that, because part of being in high school is trying things out. If pink hair is something that you need to try out for a few days, then the high school kind of overlooks that and tries to help them get over in a different way, rather than passing too many rules that they just have to break anyway, because that's what it means to be an adolescent. (Amy, 8th grade teacher)

As we can see from above, there is a huge change in the level of structure Waldorf teachers provide for high school students. While they try to control elementary school children in almost everything they do in school, they believe that at the high school level, students should have more freedom to explore their own interests, make decisions on their own, and to even challenge and break rules.

> The ultimate goal is to give the authority into their own hands ... so that they can become their own authority. If they learn to obey an adult and to follow the discipline that is given to them by an adult ... it is going to become self-discipline. And it happens around the age of 14 as they are going to the high school years.... The ultimate goal is for them to take it on themselves. (Lauren, 2nd grade teacher)

According to Waldorf educators, at the high school level students are more able to make independent judgments and are more capable of critical questioning. At this level they no longer need the teacher controlling and directing their experiences. They need instead positive leadership from someone they can establish a cordial relationship with and who can help and support them in their learning and development.

It is important to remark that from the Waldorf perspective, the authority of the teacher should arise from the quality of his/her being and from the love and respect that children have for him/her as a person rather than from coercive power (Marshak, 1997, p. 44). Besides, this authority should emerge not from egocentric urges, but from a deep desire to serve something higher than themselves. See, for instance, the following comment by a teacher:

> I'm not an authority because I take the power. It is not a power thing between me and the child like "you will do it because I say," although it may sound like that sometimes. I really take the authority because I am under authority myself.... I guess there is a sense that there is something greater than ourselves ... and something that we are reaching out to. We are serving a higher master. That might sort of touch on the Christian aspect of the school, which is certainly not something we teach the children, but it is something that lives in us as a teacher: That we are serving this higher master in ourselves and striving to become a better human being. (Lauren, 2nd grade teacher)

While thinking about freedom and structure in the Waldorf curriculum, we might wonder whether Waldorf teachers have any autonomy in their teaching at all. As I asked some teachers this question, they all answered that the Waldorf framework does allow them a lot of freedom. While they do need to follow the backbone of the curriculum given by Steiner, because they don't have to adopt textbooks and their students don't have to take regular standardized tests (unless there is some kind of state requirement), they feel they can be quite flexible in what they bring into their classrooms. A teacher explains:

> I have wonderful freedom, I can bring all kinds of things that matter to me, but on the other hand I don't have total license. This could be very subjective or a teacher would teach only certain things that he found interesting. I think there is a great balance between my freedom—I have a huge amount to say what I do with the class—but I'm also called upon to pay attention to human development. There is a right time for certain things and a wrong time for certain things. So I don't want to go against the laws of child development, you see what I mean? So, I don't see that as a restriction to my freedom. I think that is enrichment, because I found it to be true that there are developmental

> stages and certain approaches are more appropriate to a third grader that wouldn't be appropriate to a seventh grader. (Mark, 4th grade teacher)

While for many of us outsiders, Waldorf teachers are just too bound to the framework given by Steiner, for them, this framework is what gives them the freedom they need to teach within safe boundaries. While to many of us, Waldorf classrooms look very much the same, for Waldorf teachers and many of the students and alumni who I talked to, Waldorf classrooms do differ a lot. It all depends on who the teacher and the students are and on what "truly lives" within each one of them.

From the Waldorf perspective, freedom is a long-term goal cultivated through structure and control. While this might seem paradoxical, the Waldorf curriculum does appear to foster both inner and outer freedom in the following ways:

- By protecting children from the overwhelming "adult world" so that children can have a true childhood.
- By giving them a form, a direction that helps them connect with something greater than just what they are caught up with for the moment such as the latest news put out by the media.
- By requiring that students be disciplined in their work, so they can develop self-control, self-discipline, and strength to stand up to their duties and achieve long-term goals later in their life.
- By promoting a rich imaginative life in children. Imagination, as we know, is the very basis for creative thinking: We need to be imaginative if we are to visualize different possibilities and come to discoveries and insights on our own.
- By protecting children from and helping them become critical of materialistic and consumerist values, which could otherwise prevent them from being able to be who they are and from following the natural course of their development.
- By focusing on spirituality and on children's inner life rather than on their external achievement such as high grades and/or "first place."

- By fostering the development of the whole child—intellectual, physical, emotional, aesthetic, spiritual, and social—and by offering them a rich curriculum that explores their different potentials and opens them up to different possibilities.
- By not allowing children to make judgment calls and decisions in the early grades. As paradoxical as this might seem, as Steiner suggests and as an alumnus commented during an interview, this, in some ways, helps them develop the habit of "feeling and listening with their whole being" before judging, which in turn allows them to make wiser decisions.

It is interesting to note that even if Waldorf education didn't aim at promoting freedom through structure, structure would still be a necessary asset for this kind of education. The curriculum is quite rich and diversified and the classes large. Without a firm schedule to organize all the different classes, and without the firm orchestration of teachers to keep students orderly moving along as a group from activity to activity and from class to class, the whole Waldorf experience would probably become chaotic.

Notes

1. This explanation was given by Steiner in a lecture at Oxford in 1922.
2. This is a pseudonym

Chapter 11

Questions and Issues

Accountability Issues

Waldorf students usually do not take academic tests until the later grades. In the case of the SWS, they start taking tests only in 6th grade as a way of getting them prepared for the standardized tests they are likely to have to take when they leave the school. Since they are not tested before then, an accurate view of their developmental and learning level depends on the teacher's observational and organizational skills.

The fact that Waldorf teachers stay with the same group of students for a long period of time has the advantage that it allows them to know their students well and gives them time to consistently work with individual children according to their personal needs. On the other hand though, in cases in which the teachers are not able to detect a learning difficulty in the child, the latter is likely to go a long way before somebody else notices the problem. The SWS is trying to prevent this problem by having a routine evaluation with all first and second grade students in aspects such as learning differences, learning styles, dominance, reading skills, motor coordination, memory skills, ability to follow directions, and behavioral problems. The specialist who does this evaluation with the children gives teachers a good picture of where each child is in terms of learning and development, and when necessary, refers her to some kind of "remedial work" or special tutoring. He/she also helps the teacher devise strategies to deal with the needs of individual children.

Waldorf elementary teachers usually do not use grade evaluations. Instead, they write reports on the progress of students at least once a year and also hold regular teacher/parent conferences twice a year. Since they walk their students step by step in their tasks, and have them do basically the same things at the same time, generally speaking, accountability doesn't seem to be as much an issue in Waldorf education as it is in schools in which children have a large extent of freedom.

Classroom Size

The size of Waldorf classrooms varies from school to school. In the USA, they usually range from 20 to 30 students. In Europe they tend to be larger (Holland, 1981). Waldorf educators believe that children should socialize with a large group of peers. According to them, small numbers tend to generate social problems (students will be more likely to get locked into social positions). Besides, they do not allow for a certain degree of richness in group activities (e.g., plays, orchestras, choirs, and games) and peer teaching and learning.

From my perspective, while large classrooms have the advantages presented above, they also present two important drawbacks. The first is that teachers can't give much attention to individual students. And this was certainly a little problematic in the case of some of the classrooms I observed. I noticed that the Waldorf teachers tend to accommodate better the needs of the students who take longer to learn than those of students who are more advanced. Some of the most advanced students actually commented to me that they found the reading, writing and math classes a little too easy for them, and said they wished the teachers gave them more complex work to do. A few students also said they found some of the physical exercises they did during morning circle too simple. These students would actually sometimes seem a little "spaced" during these activities or would keep trying more complex moves than they were asked for. For example, if the teacher asked them to throw a bean bag up in the air and clap once, they would throw it to the air and try to clap three or four times. This obviously distracted some of the other students.

As I asked some SWS teachers how they addressed the individual needs of their students considering the large size of their classrooms (around 28 to 30 students), they tended to talk about the students who were a little behind the class, and did not mention those who were more advanced. At first I thought this was the case because they believe children should not be "further rushed" into more learning. Later I found that whereas this is true, they also believe that children who are too advanced in certain aspects tend to lack in others. And it is in these lacking aspects that the Waldorf teachers tend to focus. Therefore, while to the eyes of outsiders Waldorf teachers might seem to be holding students back in certain aspects, from the Waldorf perspective, they are not. Instead, they are optimizing the use of children's energy so that they can work on other aspects that they need more help with. Generally speaking, Waldorf educators are more concerned with addressing children's weaknesses than with promoting their special talents or aspects where they already have a good "head start." Holland (1981) explains:

> No one should ever seek to deny a child's special gifts, or to discourage a child in the pursuit of special talents. Nonetheless, there is ample time in the young person's life for specialized concentration on a narrow field of interest, and it can be done at the appropriate time in the beneficial context of a more well-rounded individual. Thus, the teacher often strives to develop expression in those areas in which the child is weaker. In fact, the teacher should rightly be less concerned in the earliest years with emphasizing special talents than with encouraging expressions of other facts of the child's being. (p. 10)

The second main problem of the large size of Waldorf classrooms is "waiting." Waldorf students sometimes have to wait a long time to get some individual help or to try a specific move that demands the teacher's supervision. As I observed, this would generate a level of discomfort among students, and sometimes, also behavioral problems. A strategy that Lauren, the second grade teacher, used to minimize this issue was to have the students who finished their assignments before others, help their peers in need of support. She also experimented using the help of an assistant teacher who would come into her classroom at different points of the day. However, given the size of the group, these strategies were not always enough to fully address the problem.

According to Ellen (3rd grade teacher), there were three main strategies she and other Waldorf teachers used to help them deal with the large group of students they had: First, they tried to meditate on the individual children on a daily basis, which helped them keep a degree of awareness of their unique needs. Second, they counted on the help of parents who volunteered to help them. Third, they tried to support each other on their work as much as possible (e.g., as I have previously mentioned, every week during the faculty meeting, they set apart some time to discuss issues they were dealing with in their daily practice). An important factor in Waldorf education that potentially minimizes the problems which are common to large classrooms is that the same teacher stays with the same group of students for a long period of time. This allows them to get to know students well, which facilitates their work in terms of dealing with their specific needs.

Technology in the Curriculum

The Waldorf curriculum doesn't focus on the teaching of technological skills until high school. This is so not because Waldorf educators are against technology, but because they believe "there is a right time for everything in life." They are of the opinion that the introduction to the use of computer and other kinds of technology at an early age will inevitably take away from children's imaginative and feeling life and will prevent them from having concrete experiences with the world around them. From the Waldorf perspective, children should only use technological tools when they are able to understand how these tools work and how they can use the latter without being used by them.

Sheltering Children

As I have previously discussed, the Waldorf curriculum tends to shelter children from the world around them. In this context, we may question whether Waldorf students will ever develop a critical view of the world and commit themselves to help change it for the better. My own view on this issue is that a sense of responsibility for the world should not come exclusively from "head knowledge," but from a feeling of deep connection with everyone and everything in the universe. As we know from experience, activism based only on "head knowledge" tends to be ineffective and is likely to lead either to violence or beautiful-but-empty discourses. While Waldorf students might seem a little alienated from the world at the level of information, they seem to be quite connected to it at the "feeling" level. As we know, this kind of connection that comes from the heart tends to be powerful. I personally believe that it has a big potential not only to move people towards peaceful and respectful relationship with others and with the natural environment, but also towards working for a better world. It would be interesting to do some research in this area to verify to what extent this really happens.

Does Waldorf Education Work for Every Child?

Waldorf education is meant for group rather than individual learning. Therefore, it is not the best kind of education for children who need individualized attention or extra help. This point is rather clear in the SWS handbook: "*The school provides a well-rounded classical education for children who are able to learn in a group setting. Custom tailored, individual instruction is not provided by the school.*" A teacher explains:

> There is not a lot of time for one-on-one attention. There is not a lot of time for the teachers to sit down to give to one or two students like OK, let's go over this math problem to see if you understand it.... In the upper grade there is a little bit more time, but by then, you know, it is too late. So everybody is doing everything together and that is not for everyone. Some children just don't learn that way. (Rose, 6th grade teacher)

According to my observations, Waldorf education also tends not to work well for children whose learning styles are self-directed and who do not like to be herded as a group. In order to do well in a Waldorf school, the student needs to be able to follow directions and to work and learn with others in a timely fashion (I will further discuss this issue in the final chapter).

Families who choose Waldorf education for their children need to be certain that this kind of education is right for them so that they will not succumb to criticisms of people who believe that the Waldorf philosophy is just "too radical," given the kinds of commitment it imposes on families. Also, and very importantly, the family should provide the child with a healthy lifestyle at home and protect her from overflows of information, TV, and other electronic media. When home-life contradicts with the school's values, the child is likely not to do well in school (e.g., children who are constantly exposed to TV and its fast-moving and over-exciting images tend to be bored with the rhythm of the Waldorf School life).

How Do Waldorf Students and Graduates Adapt in Other Kinds of Schools? How Are They Different from Students and Graduates from Other Schools?

According to Waldorf statistics, Waldorf students tend to do well in other kinds of schools and in college. They are only likely to face difficulties if transferred to different schools while still in first, second, or third grade. See, for example, the following passage from the SWS handbook:

> The order of our teaching in the first three grades is quite different from that of most public and private schools. For this reason, transferring to another kind of school during these years may prove difficult for some children. A child who has successfully completed the academic work of the third through the fifth grades can be expected to succeed in any school after a short period of adjustment. The curriculum of our upper grades meets, and often exceeds, the standards of other schools.

Throughout my literature review on Waldorf Education, I found only one independent study addressing the long-term effects of Waldorf schooling. This study was conducted in Germany and financed by the Bonn Department of Education. It included 1,460 former Waldorf students born in the years of 1946 and 1947, and was published in *Der Spiegel*, which is a weekly magazine analogous to *Newsweek* or *Time*. According to this study (Beste Einsichten, 1981), Waldorf graduates did almost three times better in the Abitur[1] tests than students from the State schools. This study also came to the following conclusions:

> The Waldorf students showed a preference for occupations in the educational and social fields (20 percent), in the medical (12 percent), and in the artistic/linguistic field (12 percent); legal and technical professions were "underrepresented." The graduates obviously took their incentives for professional choice from the Waldorf values. Success, prestige, recognition and career potential, and income played at best a subordinate role. As "personally especially important" in making their decision, the graduates named above all their own inclinations and abilities, independence and interest; then followed social and altruistic aspects. (Translated by Renate Field in a Waldorf handout)

While my own research didn't focus on the results of Waldorf education, I had the opportunity to conduct an in-depth interview with one Waldorf graduate whom I will call here Adam. According to him, some of the main gifts of Waldorf education in his own life were: (1) It gave him a well-rounded education which covered not only all the regular academic subjects, but also different artistic fields as well as astronomy, mythology, philosophy, and comparative religion. (2) It taught him to see connections between different subjects and different things in life. (3) It helped him become independent and less susceptible to different kinds of pressure from peers and from the media. (4) It fostered in him a love for learning (he never had to study for grades). (5) It nurtured in him a feeling that whatever he were to do in life, it should somehow contribute to his personal growth. (6) It nourished in him a consciousness that learning is a life-long process. (7) It fostered in him a sense of openness to people, experiences, and to what life has to offer. (8) It gave him tools to become a good critical thinker. According to him, Waldorf education helped him develop a sense of openness as it taught him to listen with his whole being before forming judgments and taking action. Also according to him, the form and structure of the Waldorf curriculum paradoxically gave him tools to become a critical thinker since he learned to "listen with his whole body," consider different viewpoints, and hold judgements before coming to conclusions about specific matters. Obviously much of what Adam sees as results of Waldorf education might also have been nurtured through experiences he had outside school. Besides, his perspective is only one perspective and cannot be generalized.

But since he did bring up all the points above, I decided to mention them here as a reference that can be used for further investigation about the benefits of Waldorf education.

Some of the things Adam feels he missed in Waldorf education were that it didn't prepare him to take tests and to deal with competitive environments such as that which he found in the university he attended. Besides, he said, he never really learned computer skills. According to some of the teachers I interviewed, Waldorf schools are now starting to address the concerns brought up by Adam. Many Waldorf high schools are now directly teaching their students technological skills and drilling them on test taking.

Curriculum Structure and Issues of Time and Place

One of the main criticisms against the Waldorf curriculum is that it is Eurocentric and outdated. Waldorf educators argue, however, that while it is true that the curriculum was created in Germany in 1919, because it was so thoroughly thought out, it is still current and still addresses children's developmental needs. They also argue that, while they do have to follow the curricular backbone given by Steiner, they still have freedom to address local and current issues according to the personal interests and needs of the particular group of students they are working with. Regardless of the extent of freedom Waldorf teachers in fact have in their daily practice, according to my observations, they did not seem to be as much concerned with implementing local and current issues in the curriculum as they were with fostering goals that normally transcend issues of time and place such as promoting the development of self-knowledge, mindfulness, and a sense of awe and reverence for the things that make life possible and worthwhile.

While I can see the beauty of this approach, along the same lines as Atsuhiko Yoshida (1999, 2000), I believe that we as teachers need to take into consideration the roots of our own culture instead of importing external ones. We also need to incorporate these roots into the curriculum through stories, tales, celebrations, folk songs and dance, literature, arts and crafts, and rituals. Yoshida, for example, is involved both as a researcher and as a parent in a Waldorf School in Japan where the whole school community is trying to adapt the Western perspective of the Waldorf curriculum to Eastern cultural and spiritual traditions. In doing so, the members of the community are not emphasizing the Christian background of the Waldorf approach. Instead, they are bringing into the curriculum different aspects of the traditional Japanese culture such as the tea ceremony, Japanese calligraphy, and origami. These kinds of activities, according to the school community, can provide the same kind of structure and form suggested by Steiner to promote the more transcendental goals which are part of the Waldorf educational philosophy. However, because the curriculum now is taking into consideration the students' cultural roots, it is making more sense to them. Besides, it is contributing to fostering in children a sense of appreciation for and a deeper understanding of their own culture and background. It is important to remark here that while I have just quoted an example of a Waldorf School in Japan, where the culture is more homogeneous than that of many other countries, I believe that education should not be rooted in one—and only one culture—in multicultural societies. Otherwise, children will not learn to respect and value the richness of cultural diversity.

Note

1. An equivalent test in the USA would entitle students to skip introductory courses and start college as sophomores.

The
Bellwether School

Chapter 12

Introducing the Bellwether School

> The purpose of the Bellwether School is to support children in developing a caring and fulfilling relationship to the world around them. We aim to nourish the healthy development of each child's distinctive abilities and personal identity within a safe, nurturing, encouraging, collaborative community of teachers, parents and children. (School flier)

The Bellwether School (BWS) was started in 1992, in Burlington[1], Vermont, as an in-home preschool by Beth Stadtlander, an early childhood educator strongly committed to constructivist and developmentalist perspectives. A school flier explains the name Bellwether: "The word *Bellwether*, in days of old, referred to 'the sheep who wore a bell and led the flock.' In modern times, the word has come to mean 'an indicator of future trends' or 'a concept or person that takes the lead.' "

Beth Stadtlander's program served a handful of children, including Justin Lloyd-Miller, the oldest child of Ron Miller, who is one of the main experts on the holistic education movement worldwide. As Justin grew older, Ron and his wife started thinking about educational options for him when he reached school age. As they didn't find any school that fully satisfied them in the Burlington area where they lived, they decided to take this as an opportunity to establish a holistic school that embraced the kinds of ideas Ron Miller defended as a philosopher and advocate of holistic education. Since they liked Beth Stadtlander's approach and found that her educational philosophy and practices were naturally attuned to the holistic framework, they offered to help her grow her preschool and create an elementary program. Ron himself tells the story:

> Beth was enthusiastic about expanding her program into a fully developed elementary school, so during the spring and summer of 1994 we began meeting with friends and colleagues in the area to discuss the project. I invited colleagues from out of state, who had particular areas of expertise in school design and holistic curriculum development, to share their thoughts with us. By the fall we had determined the shape the school would take: We decided, among other things, that it would *not* be a parent co-op, subject to the changing expectations and demands of families, but would be guided by a board of directors that included educators committed to a progressive/holistic approach. From that point, we divided our tasks, with Beth formulating school policies and educational plans, and hiring a staff, while I oversaw purchasing a site, renovating/constructing a building, and getting all the legal documents to convert Beth's school into a nonprofit organization. We were fortunate to have access to private funding that enabled us to make our dream a reality.
>
> We worked closely with an architect and a team of educational and design consultants to create an environment that would enhance the school's mission. In the spring of 1995, we began construction and started to hold open houses for interested families. Beth's little school had already acquired a sparkling reputation (and there were always many parents looking for a quality preschool), so we had no trouble filling the expanded preschool program. By the fall, when the school opened, we also had 6 children, including my son, in the new kindergarten/first grade classroom.
>
> For the first two years, Beth served as school director while teaching full time. Because our educational philosophies were so closely aligned, I stepped back contentedly and let her hire and guide the

> teaching staff. The school thrived and grew. However, in 1997, she chose to retire in order to start her own family, and I was asked by remaining staff members to take a more active role as president of the board. With the school growing and new teachers on board, the school needed more hands-on management than any of the teachers could provide. We hired parents to manage the office part time. I worked as acting director for parts of two years (reluctantly, because I have no training or particular skill as a manager, but the school community respected my role as co-founder and educational visionary). Finally, by 1999 we had hired a professional, full-time office manager and in 2000 had grown enough to promote one of our teachers to "administrative coordinator." (Ron Miller, personal communication, 2000)

During the time frame of my data collection (January to June of 2000), the BWS was run by the teachers and by a board of directors (it still did not have an administrative coordinator). The teachers made most administrative and pedagogical decisions, including hiring new staff. The board served mostly to oversee the development of the school, deal with the budget, and work as a liaison between the school and the external community.

At that time, the school had a preschool classroom with 15 children ages 3 to 5; a primary classroom with 18 children 5, 6, and 7 years old; and an elementary classroom with 9 students of 8, 9, 10, and 11 years old. In September of 2000 the school was expanded to accommodate the needs of the continuing students up to the 6th grade. The monthly fee for the preschool, primary, and elementary classrooms five days a week from 8:00 a.m. to 2:30 p.m. was $534.[2] Those who attended the after-school program paid an additional fee.

The school building looked more like a house than a school and had large and sunny rooms. Both the preschool and primary classrooms were designed in the form of the letter "L" and had many different working areas. The primary classroom, for instance, had a meeting area (a central space in the classroom with a big round carpet on the floor where children sat for meetings, group reading, story-telling, and the development of projects); a reading area with a shelf full of books, a rug and pillows; two block-building areas (one with medium size blocks and one with large blocks that could be used to build structures such as houses that would fit many children in); a wood-working area (this area had materials such as saws, hammers, nails, screws, and protective eye-glasses); a loft with many pillows for children to play and hang-out; a message center; a dress-up area; a sand-table; and an area with various tables which were arranged on a daily basis with rotating activities (e.g., cutting and gluing, painting, sewing, drawing, and modeling with clay). The elementary classroom was relatively smaller than the other classrooms, since it had been improvised to accommodate the aging group of students until the new school addition was ready. But in spite of being smaller, like the other rooms, it was sunny and nicely decorated with some of the children's artwork.

Notes

1. In the fall of 1995 the school moved to Williston, VT.
2. Monthly tuition is based on a 10-month school year.

Chapter 13

The Emergent Curriculum

The Bellwether School curriculum is "emergent" and child centered. In other words, while the teachers have specific goals in mind, the curriculum is based on subjects students demonstrate interest in. Based on what happens in the classroom, the teachers develop activities and prepare the environment in ways that connect with children's emerging interests and can potentially promote their learning and development.

Let me give a concrete example of what the BWS means by "emergent curriculum." When I was observing the school, the primary classroom (children 5 to 7 years old) was working on a project about Ancient Egypt. This project started when one of the teachers read a book to the children that talked about ancient Egyptian history and culture. As this group of students showed a strong interest in the topic, the teachers brought in more books and materials so that the children could further explore it. Since the interest in the subject spread among all the students in the classroom, the study of Ancient Egypt became a major thematic project. The class developed activities such as making arts and crafts; building up pyramids using blocks, cardboard, paper, and sugar cubes; and constructing a real person's size mummy and sarcophagus with cardboard paper. They also organized a slide show about Egypt, visited a museum to see a real mummy, and read books and wrote essays on the subject. The project lasted for months, and only came to an end as the group's interest on the theme became less intense and the children started to focus on other subjects.

The BWS teachers create opportunities for children to express themselves and give them time to pursue issues that are relevant to them. They want their students to develop a sense that their interests and viewpoints are valid and important. They also want children to realize that they have a voice, that their voice can and should be heard, and that it can make a difference in the world. See, for instance, the following quote by Andy, one of the primary classroom teachers. While he is referring to a specific situation that happened in his classroom, his words well elucidate this point:

> The children noticed the ants on the ground and it was their passion for yesterday and today, and who knows for how much longer.... They came in ready to spend an hour and a half searching for ants. They went to the preschool and they went all around the school. I think of that as a step in developing that sense of self, to be able to say, your thoughts are OK, your ideas are OK. Maybe going around and looking for ants would not be my first choice of things to do in the day. But the fact that you are getting me or giving me the time to do that, and then we are giving you the voice to explain why you want to do that, and how you want to do that, and what are the reasons that you want to do that.... I think that is a great first step for children to realize "that is my voice and my voice has strength and I have power," and further down the road it becomes ... some sort of activism, you go into work, to volunteer whatever, in a food-shelter, it starts with simple respect and voice. Respect for the voice of children and then also actively seeking that out in simple everyday things ... and giving them time. Time is the big thing. How you use it. What you are willing to give time to and to take away from? ... [I want them to] develop this inner power that they know that their voice means something and they know that they have something to say and they need to be heard....

Everything that happens at the BWS can potentially become a part of the curriculum. While I was observing, for example, the teachers, some architects, and a committee of parents had been planning the expansion of the school facilities. At a certain point in their work they decided to include the children in this process. The architects visited the classrooms, talked about their work and asked children's input on how to go about the school reform. Following the architects' visit, the teachers led many discussions about the topic and put out materials so that the students could work on the ideas they were having about the expansion. They made, for example, some copies of the architects' design for the new building and taped them on walls and tables. They also distributed paper and pencils on tables so that the children could work on the architects' design or create their own sketches. This experience was important for the students not only in the sense that they felt included in an important decision process, but also because it was an excellent opportunity for them to work on concrete skills such as perspective and three-dimensional drawing.

Note that whereas the curriculum is based on things children are interested in, in reality, the BWS teachers will only promote specific topics as major thematic projects if these topics are also personally attractive to them as adults. As they stated in interviews, they believe that they need to be excited about the kinds of things they will be working with, if they are to keep a level of interest and motivation going on in the classroom. Also note that in the case of the older grades, while the BWS curriculum is still emergent and based on subjects children are interested in, it is more structured than it is in the earlier grades so that the teacher(s) can make sure to cover different skills and curriculum areas (e.g., history, math, science, language arts). But while it is more structured, it still gives children a certain degree of choices.

While I was observing the school, for example, Cynthia, the elementary classroom teacher, had just implemented a science unit. She talked with the students about the importance of this unit and proposed that they develop it by working on a science fair. As the students very much liked the idea of the fair, they decided to go ahead with the project. Each student chose a scientific concept to explore through empirical experimentation and presented it to the class. Based on questions that emerged in class discussion, they deepened their research and then carried out their experiments in the primary classroom. The questions raised in this new presentation led them to further investigate their themes and get prepared for the fair, where they exhibited their experiments to the whole school community. As we can see in this example, the teacher did have a structure and specific goals in mind for her class: She wanted it to work on a science unit and study certain scientific concepts. She left open to discussion, however, the ways the class could go about the development of this unit and the study of the concepts. As everyone approved the idea of the fair, they went ahead with this specific project. When I sent Cynthia a copy of my research report about a year and a half after I first collected the data on the school, she told me, through an e-mail message, that the children so much liked the idea of the fair, that it became an integral part of the curriculum. According to her, the fair now includes the following stages:

- After selecting topics, each child researches the science concept and performs the experiment, with the goal of clarifying personal understanding and also generating questions.
- Each child presents the experiment to the class and leads the discussion of the concept....
- Parents are used as a resource.... We often ask if any parent has knowledge of the topic or resources to share.
- Questions we cannot answer are sent via the Internet to expert scientists who then write back to the children.
- The answers as they arrive are then discussed by the class. Even more questions may be generated and sent.
- As the discussions evolve, we begin to notice connections between different concepts and to list reoccurring themes and ideas across the various experiments, as well as interrelationships between ideas.
- Each child prepares artwork to explain the concept, including a write-up of his or her own personal understanding of the concept, and a clear set of directions and supplies for the experiment....
- The Science Fair is held in which the students perform their experiments for the school community, explain the topic, and take questions from the audience. (Everyone in the school, families and children from other classes, is invited. The Fair is on exhibit for two days.)

- Parents are then invited to link the concepts of the experiments by filling out a sheet prepared by the students. They are given a list of concepts the students had taken note of and explored to make pathways between the experiments, which they have observed. Students also fill out the sheet.
- In a culminating activity, we share and discuss all the diverse connections different people have made. (Cynthia, e-mail message, 2001)

Aspects of the Curriculum and School Day

The day at the BWS started smoothly. Children would arrive from 8 to 8:30 with one of their parents or another adult, who would often spend some time in the classroom interacting with his/her child, other children, and/or with the teacher. From the time they arrived until about 9:30 they had "explore time," a time when they could engage in any activity of their choice (the teachers would come in early in the morning and prepare the classroom with different materials and activities). Some examples of activities they did during explore time were knitting, embroidering, sewing, electric circuit building, jewelry-making, weaving, writing, drawing, painting, clay modeling, puppet-making, puppet shows, recorder and drum playing, block building, sand-table and water-table play, wood-work, table games (e.g., chess, checkers, shadow projection), or just talking with each other. An interesting characteristic of explore time was that as all classrooms at BWS had it at the same time, children were free to visit each other's rooms, chat with peers, play, and explore different equipment and materials together.

After explore time, they cleaned up and had a meeting. In the case of the elementary classroom, the meeting started around 9:40 and lasted until about 9:50. During this time teachers and students shared news, talked about the activities they developed during explore time, and discussed issues of interest. For example, they talked about things they did during the weekend, showed the projects they developed during explore time, or talked about things such as how they could be more supportive of each other's difficulties and welcome the accomplishments of others in the classroom. Following the meeting, the elementary classroom had a block of about 25 minutes (usually from around 9:50 to 10:15) with rotational activities. Some activities I observed during this time were the study of Spanish and geography, the practice of the recorder and spelling, and the exploration of scientific projects for the science fair (e.g. they explored the concepts of and developed experiments with electromagnetism, gravity, viscosity, flight, and water pressure).

From around 10:15 to 10:30, the elementary classroom teachers and students sat in a circle where they shared the food brought in by a volunteer parent (the parents rotated in sending in snacks for the whole group of students). From about 10:30 to 11:45, on Mondays, Wednesdays and Fridays, they had math workshops. On Tuesdays and Thursdays they had writer's workshops. The teachers would give them some assignments to work on during this time and the children would work either individually or in small groups. The teachers would then go around in the classroom helping the children who needed assistance.

After the workshops, the elementary classroom had about 45 minutes of outside play. During this time they engaged in activities of their choice such as tag, sledding, snow fort building, sandbox play, jungle gym climbing, ball games, and bird watching. At 12:30 p.m. they had lunch, and around 1:00 p.m., "shared reading time." During this time, Cynthia read aloud a piece of writing and then led a discussion about the reading. The reading usually consisted of a chapter in a novel or biography, an informational text, or a current event article.[1]

From about 1:30 to 2:15 the elementary classroom had either quiet reading or worked on thematic projects. During quiet reading the students silently read books of their choice while Cynthia worked with them on an individual basis according to their specific needs. Every Thursday afternoon Belle, a volunteer parent, taught a unit on the natural history of Vermont through the development of various artistic activities. Every Wednesday morning, the elementary classroom had a music class with a special teacher. On Friday afternoons, they had an "all-school-meeting" where they talked about things that were of interest to the whole school. Also once a week they worked on something they called "homework contract" where each child developed a plan on the kind of work they would complete at home. For this contract, Cynthia would usually give them some directions about what kinds of things they were supposed to work on, granting them options on how to go about their work (e.g., she would ask them to write an essay of a page or two, and each of them would choose the subjects they wanted to write about).[2] The regular school day at the BWS ended at 2:30. Those who stayed for the after-school program remained in school until the end of the afternoon.

The primary classroom had a similar schedule with a few variations: They started with explore time followed by clean-up time until about 9:45. They then had a meeting from about 9:45 to 10:05 when the whole group got together. After that they had about 25 minutes for snack, and at 10:30, they had either math (Mondays, Wednesdays, and Fridays) or writer's workshop (Tuesdays and Thursdays). From 11:30 to 12:15 p.m., they had outside time. From 12:15 to 12:30 p.m., they had chapter book reading, and from 12:30 to 1:00 p.m., lunch. From 1 to 1:30 p.m. (Tuesdays and Thursdays) or from 1:00 to 1:45 p.m. (on Mondays, Wednesdays, and Fridays), they had library time. During this time the children chose books they wanted to look at or read, and the teachers worked with them on an individual basis on their reading skills. On Tuesdays and Thursdays, from 1:30 to 2:15 p.m., the group had either Spanish, Science, Music, or worked on thematic projects. On Mondays and Wednesdays, from 1:45 to 2:15 p.m., they had spelling/literacy and on some occasions, music. Like the elementary classroom, they also had a special music class on Wednesdays and participated in the "all school meeting" on Fridays.

The preschool also started the day with explore time. At around 8:45 a.m., they had a brief greeting meeting where everyone had the opportunity to say hello. Explore time then continued until around 10:00 a.m., when they started to clean up. At 10:15 a.m., they had a group meeting when the teachers read books, role played, sang with them, introduced new materials, and discussed ideas or issues that were important to the group. The meeting was followed by snack time from 10:30 to 11:00 a.m. From 11:00 to about 11:30 or 11:45 a.m., they had outside time. Quite often, at around 11:30 a.m., they also had what they called "friendship meeting," which was a time of the day where the teachers mediated a conversation between the children about their relationship with peers and helped them address problems they were having with each other. Lunch started at around 11:45 and lasted until around 12:15 p.m., when the part-time children started leaving. At about 12:30 p.m., the teachers had all the children on mats and read them some stories setting the tone for quiet rest time. The children were expected either to sleep or to play quiet games from about 12:45 to 1:45. "Afternoon choice" time started at around 1:45. During this time the students would usually draw, "wrote" letters and journals, and played with puzzles and table blocks, among other things. The children who did not stay for the afternoon program left at around 2:30 p.m.

Some important characteristics of the preschool curriculum were (1) It was play and child-centered and did not rush students towards academic work. While there was a writing center and many books available in the classroom, the students would only engage in reading and writing if they spontaneously decided to do so. (2) It placed a strong emphasis on promoting children's social skills and a sense of community in the classroom. (3) It fostered autonomy and independence (this was done, for example, as it encouraged students to make decisions about most of the activities they engaged in at school and as it involved them in many practical tasks such as preparing snacks, helping themselves at the table, and cleaning up the classroom). (4) It emphasized the development of children's thinking and communication skills, which was done, for instance, as the teachers gave students choices about what to do, and as they asked them questions about what they were doing, why they were doing what they were doing, and how they were feeling about what they were doing.

It is important to note that while the BWS teachers did follow the structure of the day described above, they remained flexible to accommodate the emerging needs of the students (e.g., if there were an important conflict between the children, they would not hesitate addressing it right away). They also remained opened to adapt their schedule and activities to take advantage of what they foresaw as emerging "teachable moments." Toby (primary classroom teacher) objectively commented on this subject:

> Whenever it seems to be a good time to use math or do something about literacy while we are doing something else, we will. It's not just that we do it at literacy time, or at math time, but whenever we can, we put these things in a meaningful way to teach it to the kids.

Let me give an example of this. Once I was observing the primary classroom when Toby was starting to do the calendar with the students (this was a daily routine task in her class). She pointed at a board that had the days of the week listed, and under each of these days, a little pocket which was supposed to be filled in with the day of the month and the weather of that specific day. She started the activity by announcing that the day was May 1st, and that they needed to take apart April's calendar to start doing the new month. But before doing so, she asked the students to count how many sunny, rainy, cloudy, snowy, and mixed days (e.g., rainy/sunny, cloudy/sunny, and rainy/snowy) they had had the past month. In the sequence she asked them a series of questions such as "How many more cloudy days than snowy days have we had? " "How do you know that?" "Which kind of day did we have the least?" "How do you know that?" "How do you know what day of the week is today?" "What number is this one?" "What does it say here?" Only after a large section of questions and answers did she take the calendar apart and had a child fill it in for the new month that was starting. At the end of this activity, in order to get the children organized to wash their hands and get ready for snack, she made calls such as the following: "Whoever has a name that starts with T can go;" "Whoever has a name that starts with the sound of the word 'dog' can go;" "Whoever has a name that starts with A can go." As we can see from this example, Toby—and I would say the other teachers as well—would promptly take advantage of a routine situation such as this one just described whenever they felt it could become a valuable learning experience for the children.

Final Comments

As we think about the emerging curriculum at the BWS and the fact that it was based on themes students were interested in, a question that often comes to mind is "But what if a child or a small group of students wanted to explore a topic that the other children were not interested in?" The answer is that only the topics that most, if not all, children were interested in became major thematic projects. When only one child or a few children were interested in a subject, the teachers would usually acknowledge the child's or group's interest, talk about it with the child and/or group of students, and whenever possible, they would also bring in materials and books so that the child or group of students could further pursue their particular interest in the subject during "explore time."

Another question that tends to come to mind as we think that the BWS curriculum is based on themes students are interested in is "But what if the latter kept on changing their interests, not being able to get in depth in any of their experiences?" The BWS teachers tried to balance students' experiences in ways to have them both explore emergent interests and also pursue established goals. While it was true that sometimes the students did remain only on a superficial level in certain experiences, once goals were set, the teachers tried to keep the children focused on these goals. When Cynthia and her students, for example, decided that they would have the science fair, they did not give up the fair project until it was done. Besides, independently of the topics children were interested in and working on, Cynthia made sure to keep them regularly practicing reading, writing, and mathematical skills.

Notes

1. Some of the advantages of the shared reading time and of the discussion that followed the reading, according to Cynthia, are that they expose children to different kinds of literature, nurture their empathy with different situations people might find themselves in, expand their vocabulary and knowledge of different subjects, promote their comprehension skills, and nourish their imagination. Other advantages are that, during these activities, the students who tend to have more difficulty to express the richness and complexity of their thoughts in writing have the opportunity to express it orally. Besides, the shared reading time is, in many cases, a great springboard for the study of history and the development of art projects.
2. The amount of homework the BWS elementary classroom students were asked to do was significantly smaller than that students from other schools are usually required.

Chapter 14

A Special Light in the Eyes

Soulfulness in the Classroom

One doesn't have to spend much time at the BWS to feel an air of enthusiasm in the environment. While I was at the school, for example, I often observed teachers and students vividly working together with a smile on their faces and a special light in their eyes, as if they were really enjoying what they were doing. Throughout interviews, the teachers openly expressed a deep love for children, for their work, and for the school. Like the teachers, the students also seemed to love the school. I conducted in-depth interviews with all the children from the elementary classroom (individually and in small groups) and they all, with no exception, kept saying only positive things about the school and their experiences there. Their love for the school was in fact confirmed to me by all the parents I interviewed. See for example the following quote by one of the mothers:

> He wakes up early in the morning and he is like "is it time for school?" He gets really upset with us if we are running a little slow.... And he just hardly waits to get to school everyday.... The school breaks are very long for him, because he is waiting to go back to school. In public

> school it was totally the opposite. He would cry and say "I can't go to school, I don't want to go!"

Perhaps the best word to describe the kind of atmosphere of the BWS would be "soulfulness." This is not meant in the religious sense, but in the sense that holistic education theorist Jack Miller (personal communication, 2000) would define it: as a quality that makes of the school "a place where people want to be because of the energy they feel." According to J. Miller (2000), some of the main characteristics of soulful schools are that they: (1) acknowledge the importance of the nonverbal and recognize that "the quality of our being and presence has as much impact on students' development as anything that we say" (p. 110); (2) pay attention to the aesthetic environment; (3) tell stories about the school, which fosters a sense of shared meaning and values; (4) have celebrations and rituals, which promotes a sense of connection between communities; (5) encourage truth and authenticity; and (6) nourish voice by creating an environment where people can express themselves and speak from their heart without fear.

But what is it that made of the BWS such a special place where people wanted to be? According to my observations, while the school did have many of the characteristics mentioned above—some more than others—the three main factors that seemed to make it such a special place were: (1) the extent to which the school nourished a deep sense of community between people; (2) the ways in which the teachers nourished children's souls and fostered in them a sense of wholeness; and (3) the "qualities of being" of the teachers. In the following, I will talk a little bit about each one of these factors.

Promoting a Strong Sense of Community

The BWS teachers were quite dedicated to fostering a culture of respect, trust, and safety in the school and classrooms. They worked on their relationship with each of their students and tried to promote an environment where the children felt acknowledged, respected, and safe to be who they were and to express their thoughts, feelings, and needs without fear. They in fact structured the school curriculum in a way to promote a sense of community not only in their classrooms, but also in the whole school. During the morning meetings (which happened on a daily basis) and "all-school-meeting" (once a week) children had the opportunity to talk about and discuss in a group issues that were important to them. Also, during explore time, snack, outdoor time, and lunchtime, they were free to play and interact with each other as much as they wanted to and as long as they were not disturbing others.

The BWS teachers explicitly and directly encouraged students to express their thoughts and feelings. They tried to make sure that children listened to and respected each other's ideas. They would not hesitate stopping any of their activities to address social problems right when these problems emerged. They actually did so to such an extent that they usually would not let go even small incidents such as the following:

> It is explore time and the students are engaged in different activities. Joshua (8 years old) suddenly yells at a younger child from the primary classroom: "Go away. I want peace and quiet here!" Cynthia who is standing on the other side of the room hears Joshua, walks over to him and says: "I would like our class to be a welcoming place to everyone!" Joshua replies: "But I need peace and quiet!" Cynthia kindly answers: "But you can say that in a gentle way." Joshua stays silent and calms down. A little latter Joshua starts rudely complaining about the children from the primary classroom who, according to him, were taking their scissors away. Cynthia asks him to re-estate what he had just said in a nicer way and invites him to discuss the problem along with the whole group so that they can work out a strategy to address it.

Another good example of how the teachers dealt with conflict and promoted a sense of community in the classroom can be seen in a story told to me by Cynthia, when I asked her about situations that directly or indirectly helped her shape her educational philosophy:

> ... I am thinking of a specific event that happened in class early in the year in which one child accused another of having stolen something. And this became a huge issue at the end of the explore time, because there was denial, there was dramatization, then the class began to polarize and argue "yes the child did it," "no that child didn't." "How do you know that child did or did not?" And so we had a meeting, a dis-

cussion and talked about "how do you know that" ... and "how do you want to treat each other?" And my take on it was "if one person says that there was no stealing, then I will believe that person...." A child burst in tears after I said that and I had an inkling that it had occurred. So there was further discussion and more accusation, and that got to a point that I said, "Let's stop talking about this. Let's not talk about this particular situation right now. Let's talk about this thing from a little larger perspective." In other words, "What kind of classroom do you want to have? Do you want the kind of classroom where people can make a mistake and to be able to be brave enough to say they made a mistake?... Do you want a kind of classroom where if you make a mistake you hide and you continue to lie and are afraid because people are going to get you and to tease you?"... So we had a huge discussion about this and they said they wanted a classroom where people could make mistakes. And I said: "How would you treat a child who made the mistake?" And they agreed that there would not be teasing. Then the child who was being accused asked to go outside and talk to the one who was accusing. And as the two were out, I said to everyone else: "This is about you too. Because if they come in, can you imagine how scary it is to be out there admitting that you did something wrong and then you will walk back in and then everyone will ridicule you for it? It is up to you, what do you want to do. This is your class, how are you going to handle this when they come back in?" So there was this very huge discussion and they were so loving about it. I think that was the most beautiful moment I had ever seen in teaching.... They were so kind, when they came in one child who had been in the main group went and hugged the person and said: "I have made a mistake somehow like you....You did the right thing!" And to me that was a very spiritual moment because ... we were learning about right and wrong and then how do you trust, how you can be wrong, we are all wrong at some point ... how can you forgive, how can you welcome back in. If you weren't part of it, be a part of helping it, not hurting it. It was a huge lesson to be happening and they responded so compassionately. Now does that mean that they have never since been teasing? No, this goes on. The kids are human, we get aggravated, we fight, we make mistakes, but at that moment, they revealed what can

> happen, and it can be extraordinary.... To me that was an incredible teaching moment. We were all learning from each other and I don't think they will ever forget it.

From the statement above we can see that Cynthia—and I could say that this was true for the other BWS teachers as well—nurtured a sense of community in the school based not only on moral stands, but mainly on love and compassion for other human beings. As she and the other teachers asked their students questions such as "What kind of community do you want to build?" or "What kind of classroom do you want to be a part of?" they helped students think about and verbalize their innermost desire to build and be a part of a caring community. They did not have to impose behavioral rules. Instead, they created them with the students.

During the year when I developed my research, the preschool had a practice that they called "friendship meetings," which in the subsequent years was adopted by the other classrooms as well. These meetings happened when the teacher or a child felt there was a need to discuss an ongoing relationship issue in the classroom. The teacher guided these meetings and tried to help students communicate their ideas about the issue that was going on in honest and non-aggressive ways, being sensitive towards each other's limitations, and taking into consideration different viewpoints.

According to my observations, the BWS did feel like a true community not only in terms of students' relationships with each other and with the teachers, but also in terms of how the teachers related to each other and with the parents. They would come early every morning to prepare their classrooms, and as they did so, they visited each other's room and discussed different subjects concerning their teaching, children, and personal life. They would talk, for example, about what worked and what did not work in their latest activities, what activities their students seemed to like the most, what kinds of books they should read and get to students, or how they could help each other to solve problems they were having with specific students.

The parents were quite present in many of the school activities. As I have previously mentioned, they actually walked their children into the classroom on a daily basis and many of them ended up spending a considerable amount of time in the classroom. Also at the end of the day, when they came in to get their children, they would often stay for a while talking with other parents, with the teachers, and with the children. Many of them volunteered in committees that helped raise funds for the school, organized events to strengthen school community, promoted public relationships outside the school, maintained the school building, published the school newsletter, etc. They also volunteered in many activities such as plays, field trips, and special teaching units. Belle, one of the parents, for example, would come every Thursday afternoon to develop a unit on, and an art project around, the natural history of Vermont. An interesting characteristic of the BWS was that the parents were welcomed in the classrooms at any point during the school day and were often encouraged to share some of their knowledge and experience in the classroom with the teachers and students.

Nourishing the Soul and Fostering a Sense of Wholeness

Rachael Kessler (2000) talks about ways in which secular schools may nourish children's souls. From her perspective, schools can do so by honoring children's search for meaning and purpose; by creating a safe environment where they can explore their questions and where their questions are honored; by offering them opportunities for silence and solitude; by honoring their urge for transcendence (this can be done, for instance, through the arts, sports, academics); by honoring their hunger for joy and delight (which can be done, for example, through plays and celebrations); by giving children opportunities to explore their creative drives; and by attending to their need for initiation.

According to my observations, the BWS experience nourished a student's soul in many of the ways suggested both by J. Miller (2000) and Kessler (2000). For example, both explore and outside play time allowed children a lot of time for free play and exploration, which directly and indirectly addressed what Kessler says about honoring children's "hunger for joy and delight," " urge for transcendence," and "expression of their creative drive." Besides, explore and outdoor time allowed students to work with their own questions and to be silent and by themselves, if they chose to do so. Another example is the way the curriculum was structured allowed children to explore things they were interested in. And as the teachers supported them in their individual pursuits, they honored, in many ways, children's search for meaning and purpose. Also the teachers constantly created opportunities for children to talk about things that were important to them and to discuss issues that needed to be addressed by the whole group. The kind of environment they created allowed children to talk "straight from heart, and without fear," as Jack Miller would probably put it.

Some other ways in which the BWS environment, and more specifically the BWS teachers, nourished students' souls were, for example, by accepting children for whom they were as individuals and by promoting their self-esteem through the acknowledgement of their personal strengths. For instance, if they found that a child was not a good writer, but could tell a story well, they would create opportunities for this child to orally tell a story so that his/her talents could be recognized. In fact, I did often observe some of the students making comments such as " I am not very good at doing this, but I am good at that," indicating that while they were aware of some of their weaknesses, they were also aware of their gifts.

The BWS teachers also nourished students' souls and a sense of wholeness as they emphasized the latter's development as whole beings and in all their aspects (intellectual, emotional, physical, social, and spiritual). They addressed, for example, the spiritual and emotional aspects of children as they accepted them for who they were as individuals, encouraged them to express their thoughts and feelings, and supported them as they addressed personal and group problems. They also fostered the emotional self as they connected the curriculum with subjects children were interested in, increasing in this way their motivation to learn. They addressed the physical self, to the extent that almost all school activities involved hands-on work and to the extent that children had a lot of time for free play both during explore time and during outside time. They fostered the social aspect, as they promoted, for instance, a significant amount of group work and collective problem solving. They integrated the intellectual, emotional, spiritual, and physical aspects as they granted children freedom of movement. They allowed, for example, that children freely moved around in the classroom as long as they properly engaged in the proposed activities and were not disturbing the class. They also allowed students to do activities such as knitting, crocheting, and drawing throughout classes and meetings when they didn't need to use their hands for specific activities. This freedom of movement, according to some of the older students who I interviewed, contributed to their sense of well being since it helped them dissipate some of their physical energy and calm down.

The kind of environment created at the BWS nourished a sense of wholeness also in the teachers. While their wages were far from ideal, they felt grateful and happy for being a part of the school community and for being able to help make of it the kind of school it was. They felt respected since their voices were not only considered in all decision-making processes, but had a determining role in the school's administration, philosophy, and practices. They also felt respected because they had complete autonomy to go about the development of the curriculum. According to the teachers themselves, they felt at home in the school since they could be very much themselves there and as they educated children, they could also work on their own education.

The Quality of Being of the Teachers

Many authors, including Ba & Josette Luvmor (personal communication, 2000); Scott Forbes (personal communication, 2000); Kessler (2000); Krishnamurti (1981); Levy (1996); J. Miller (1993, 1994); R. Miller (1995); Palmer (1993, 1998); and Steiner (1995), among others, address the importance of the "quality of being" of teachers. According to them, who the teacher is and the kind of relationship he or she is able to establish with students is one of the most, if not the most, important factor for an effective education. My research came to confirm this viewpoint, and in this section I will discuss the qualities of being of Cynthia, one of the BWS teachers. While all the BWS teachers deserve a section here, I will develop a case study only on Cynthia, in whose classroom I spent most of my time and whose students I interviewed. By focusing only on her case, I will be able to get more in depth in the discussion and offer a more detailed and well-grounded description of how the qualities of a teacher can impact the sense of well being, wholeness, and growth in the classroom.

Cynthia attended graduate school in English literature and taught English at the college level. After she had children, however, she found herself in love with children and decided to go back to school and study to become a schoolteacher. Cynthia's first contact with the BWS was as a parent. As she really liked the school's educational philosophy and practice, she joined the school staff. She taught two years collaboratively with another teacher, Toby, in the primary classroom. At the time of my research, she was completing her second year of teaching in the elementary classroom, where she was the sole teacher.

During one interview Cynthia elaborated on some of her main goals for her students: (1) That they develop a solid and strong sense of self; in other words, that as they get to know the world around them, they become aware of who they are, how they think, how they learn, what they like, where they came from, where they want to go to, and what they wish to accomplish in their life. (2) That they become creative, caring, and compassionate people; that is, people who can think for themselves and can accept others for who they are, people who are skillful in solving problems and in working with others regardless of their weaknesses and differences, and people who are able to take responsibility for their own actions. (3) That they develop strong learning habits and academic skills, so they can become life-long learners and achieve the goals they set for themselves in life. (4) That they always do the best they can in whatever they propose themselves to do. (5) That they "develop joy and wonderment in living and learning." (6) That they cultivate loving and stable relationships. (7) That they come to do some kind of interesting and meaningful work in their lives.

I have interviewed all Cynthia's students, individually and/or in small groups, and all of them, without exception, stated that she is a wonderful teacher. According to them, some of the reasons why they like her so much are that she is fair; she gives attention to each one of them; she always tries to do her best to meet their needs; she gives them "the right kind and amount of work," which in turn "makes learning effective and fun"; she demonstrates to like them; she is kind, caring, forgiving, and playful; she is open and flexible; "she considers how hard one works and not necessarily how much one gets accomplished in the day"; and as one of the students said, "she can always make you feel good when you have a problem." It is interesting to note that nine out of nine students talked about the four first qualities just mentioned, suggesting that these were quite visible to them.

Like the students, all the parents I interviewed also expressed their admiration for Cynthia for the same reasons as their children. But they also talked about two other qualities not yet mentioned: Her ability to honor each child for who she is, regardless of her weaknesses, and her capacity to communicate and to maintain a good relationship both with them and with their children not only in easy times but also in "hard times" when the latter were having difficulties. My observations confirmed all the qualities both parents and students attributed to her. In the next sections, I will focus on some of her virtues that were most evident to me.

Looking with Loving Eyes
Seeing the Beauty, Trusting the Goodness

According to Ron Miller (personal communication, 1998), one of the main characteristics of holistic educators is that they believe in the wisdom of human nature and in the "goodness" of children. They believe that children naturally want to learn, and will learn if given the opportunity to learn. Cynthia seemed not only to believe in the wisdom and "goodness" of children, but by the way she talked about them—and I would say also by the way she actually treated them—she seemed to like them all, regardless of their weaknesses and problems. See for instance two passages by her from two different interviews:

> Children want to learn. They might not always be interested in whatever we are doing at the moment, and for me then I have to look at it and try to determine why that is. Is there something that they are not understanding? Is there something that is going on? Basically children do want to learn. They're curious, and they are joyous, and they want to learn what interests them, we all do. And so I try to have a lot of choices in the class.

> I have this belief that these children are wonderful, they are here to be loved and appreciated particularly when things are hard.... I like each child and I think they know that. I think that this makes a good connection between us. I also trust them, I really trust that they won't cross boundaries that shouldn't be crossed. I really believe they just won't do that.

Cynthia would normally think the best she could of each child. She believed that if a child presented behavioral problems, it was not because she was bad, but because something was going wrong in her life that was leading her to the problem. Cynthia was also patient, compassionate, and non-judgmental. She tried to understand and meet each child according to her needs regardless of the difficulties the latter was going through. She also tried to make sure no one was excluded from the group.

> I try as much as it is humanly possible to think the best of the situation and not put a negative judgment on something. If there is a difficulty here, it's not that the child is bad, there is a difficulty here for us to work out and so we are going to try to understand it, and there is always a positive way to grow from it. It doesn't have to be a tragedy. I want, as much as I can, to be patient, to try to understand what is going on, to very calmly ask them, require them to participate with me.... I am always learning from them, but I am also always asking 'how can I teach them?' As I meet with them one by one, I ask "How can I be your teacher?" "Who are you?" "What is it that you need?" "What is it that I can get to you?" And I pray I can give it, I pray I have the wisdom and a moment to respond in the best way I possibly could respond, and I won't always. But to me that's part of the spirituality of it: That nobody is cut off, nobody is excluded. That they don't look with impatience or judgment. I want to look with loving eyes at what is going on. It is this constant process of trying to pull people back in, pull them back together, and try to reflect back over and over the wonderfulness of who they are!

There was a situation at the school during the semester I was observing that well illustrates some of Cynthia's qualities just mentioned.

Many children from different classrooms had built a snow fort where everyone played during outdoor time. A given day the fort was found on the ground. Those who used to play there felt really frustrated and angry at what had happened and wanted to know who had knocked it down. A parent, Belle, who came in to volunteer-teach a class, tells what happened right before her class:

> The kids had just come in from recess and they were having lunch and everyone was saying "I hate her, she knocked over the snow fort." ... They had had an all school meeting about it in the morning ... and had decided that these things happen and you just have to go on. But still after recess there was this lingering "we have to find out who did it!" Cynthia was so good. She just sat with the kids and said "There seems to be some real feeling of turmoil here and anger and being upset" and she said "how are we going to sort it out?" One of the kids said "we have to find out who did it!" and Cynthia said very sweetly: "Suppose you do find out who did it, what would you do then?" "Well, we have to find out who did that!" and somebody else said "I just hate her!" and Cynthia said "how do you know this person in the other class knocked down the snow fort?" Someone else said "I heard so and so say that she did!" and Cynthia said "You didn't see this?" "No I heard somebody say."... Cynthia was just so calm about that. It was just another problem-solving situation. And then in about 15 or 20 minutes, everyone had their peace and calmed down and I thought it was remarkable that she could just push aside what was going on and a kind of give a calmness to the whole thing. She was not saying you shouldn't be angry and upset. She was saying what do you want to do about it? And really getting them to say: "So a little kid did something, so what? You go on!" Now of course I was upset because Diane [Belle's daughter] was one of the kids who was most vocal in saying "this is terrible, I hate her, look at what she did!" and I thought "Oh, my sweet daughter, I never heard her talking like that!" You know here we encourage her through all her life to be loving and giving and here is my daughter and I said to Cynthia "I can't believe she was saying this" and Cynthia said "Look, kids are focused in the moment ..." so I thought it was such a gift that she could help them see what the prob-

> lem was and then go on and go by it. When science started I was the one in an emotional wreck, but Cynthia was fine and the kids were fine....

Cynthia could have not allowed her students to discuss the snow fort knockdown issue in the afternoon, since they had already talked about it in the morning. However, as she realized that this matter was still burning for some children, she gave them time and support to address it. She didn't judge and accuse children for being too picky or aggressive. Instead, she looked at them with "loving eyes" and sensibly helped them deal with the problem in a healthy manner.

Presence and Openness

Cynthia seemed to be able to fully live in the moment in the situations she found herself in. When she was with someone, she was with this someone as if there was nothing else more important for her to do. This made each of her interactions with everyone in the school a special circumstance, a moment where people felt heard and cared for. A parent explains:

> She is a remarkably good listener ... and you need to be a good listener in order to really understand what the other person is saying. On the surface people can speak words and they sound simple and understandable, but you really need to listen to the context and the history and the hopes of the person speaking. Cynthia is intuitively very empathetic towards all sorts of ages.... She is very good at making the moment the most important part of that child's life, all the children together in her life with them.

As Cynthia could be very much present in the moment with the people she was with, she was also flexible and open to new ideas. During the year I collected the data for my research, for example, a family who was planning to apply for the school started a dialogue with her to have their child, Diane, attend the school part time (only a few days a week). Most teachers would not even consider this kind of situation, since this would be likely to break the natural rhythm and dynamics of the classroom. Cynthia, however, fully discussed the situation with the family and considered their plans. As she talked with the family, she came to the conclusion that coming to school only part time would be beneficial to the child in two major ways: Even though she would miss some of the school experience, she would still be educated by her parents who were fully committed and able to take good care of her education at home. Besides, and very importantly, she would be able to keep working as a volunteer in a cooperative organic farm planting and distributing food for low-income families (she had been working on the farm since a very young age). Cynthia also found that the child's experience outside school was positive not only for Diane, her family, and for the people who she helped with her volunteer work, but also to everyone in the school to the extent that she would share her experiences with them.

A different situation that well illustrates Cynthia's openness to listen to, to take into consideration, and to address the needs of individual children and their families is exemplified by the following passage by a mother during one of my interviews:

> Homework has been an issue for Charles and James and I. We have difficulties on how to approach it. James is like: "Here is how the rules are" ... and I am more: "You know, this is an issue, let's figure out what is going on." James came to Cynthia and said that Charles wasn't being responsible, pro-active enough, and so on. He came to Cynthia and presented the solution without talking too much about the problem. So Cynthia said OK, let me think about this and let's talk about that. So she drew out of James what his concern was, and she said "do you mind if we bring Charles into this because we are talking about him?" The three of them had a little meeting to talk about why homework was important, what Charles was going to get from it, and what he thought about James' feelings. Charles was able to say, you know, "I do realize that this is important, and I haven't been doing my homework when you asked me to do it. I can work on that." That was a somewhat of an academic issue, but it was also related to our family dynamics. So these are the kinds of things that Cynthia will do instead of just saying: "He has to do homework and you figure out how to make it work," She doesn't take that approach at all.

In this example Cynthia listened both to the father and to the child and mediated a conversation between them. Because she was present both to the needs of the father and of the child, she was able to help them effectively and amicably address the problem. This kind of attitude of listening to different parts and helping them negotiate a solution to problems was very much a part of her way of being.

Calm, Yet Full of Spirit

Cynthia avoided exposing her students to criticisms. Usually, when problems came up, such as when a student would get too noisy or restless in the group, she would gently talk with the child and ask him or her to calm down. Sometimes she would also discretely call the child for a private conversation and explain that what she was doing didn't seem right. She usually remained so calm during these conversations that even though they often occurred in the middle of the classroom where everyone was working, many of the students wouldn't even notice they were going on. Cynthia was consciously understanding, gentle, and non-punitive:

> I often have to tell them, in little ways, to settle their bodies or not to hit or do some things. But I try to do it in a way where it's not punitive, but saying, 'you know, that's not the kind of behavior we would like,' and to gently bring them to a place where I want them to be, where they know it's right and want to be, but not let them feel that they are a bad person … we all make mistakes.… I also try to acknowledge, hopefully, more positive aspects of who they are so that I can focus more on that and not say so much about the negatives. I think that one of the hard pieces of being a teacher … is child management in a way that's not punitive.… It's learning how to gently guide children.

Let me give an example of a situation I observed and took notes on as it happened:

> Cynthia is fixing the snack on a large cloth on the floor [they always had snacks on the floor]. As she observes that Tania had finished her assignment, she asks the girl to get some napkins for the snack. Tania purposefully drops them all on the cloth and the napkins spread all

> about. With a soft tone of voice and a smile in her mouth Cynthia comments "it would help if you spread the napkins around for each one of us." Tania smiles and fixes the napkins. Meanwhile Joshua intentionally falls off the bench where he had been lying down since he finished his assignment. Cynthia quietly comments that he needs to find something to do until everyone else finishes her work. He says he doesn't know what to do. She suggests he help Emerson in the puzzle he had been working on. He says he doesn't want to do so and throws his sweater on the floor. She asks him to put it away, which he does. Suddenly he starts behaving well again as if nothing had happened. Cynthia spoke so softly to both Tania and Joshua that nobody seemed to notice what was going on.

The situation just described could have been completely different had Cynthia yelled at her students. However, because Cynthia was so calm and gentle, there were no heavy feelings involved between her and the children. But whereas Cynthia was calm and gentle, she was also firm when she felt certain behaviors were not appropriate or when she thought students had to complete specific tasks. Once for instance, she was reading a chapter of a book for the group when Joshua, who had been acting silly for a while, suddenly shouted, scaring everyone. Cynthia stopped reading and told him that his behavior felt hurtful to her. Joshua stood up and announced that he was "going to sit beside the lap-top girl" (the "lap-top girl" was me taking notes of my observations). Cynthia gently, yet firmly stated that she just could not keep reading while that kind of behavior was going on. By doing so, she did not directly punish him. But as she stopped reading, she led him to feel the "weight" of his behavior. He knew that the "chapter-book reading time" was a very special time of the day for everyone in the classroom and that he had probably hurt everyone's feelings for causing it to end earlier that day. He behaved better till the end of the afternoon.

"Cynthia is calm, yet, full of spirit" is an expression one of her students, Diane, used to describe her, a viewpoint which I found accurate for all the reasons I brought up above. She is full of spirit since she has a passion for life, for her work, and for her students and is able to be fully present in the moment with the people she is with. And yet, she doesn't overwhelm the latter with her presence and passion and seems not to lose control of her temper in face of difficult situations.

The Art of Being Fair and a Good Facilitator

Cynthia was fair in terms of sharing her time and energy with all students.[1] When Cynthia was with a child, she would be with that child. She would normally not stop working with this child because somebody else was not doing the required assignment or was misbehaving, unless this were extremely necessary (e.g., if a student were doing something unsafe or disturbing other children). During one of my interviews, I asked her how she balanced her time and energy with her students. She explained:

> Sometimes it is a matter of saying, "OK these two people are going to goof off at this time."... I need to honor this other child that I am with right now. This child needs my attention right now and so I am not going to get into it.... Later I follow up with them.... I feel that it is not right, always to be focusing on those kids who are just trying to pull me around and play with me a little bit and have fun, and disrupt everybody else. I mean, sometimes I feel that the other children need a chance to focus and my attention just as much as the other one does....

Besides being fair, Cynthia was also a good discussion leader. She would always try to make sure everyone had an equal share of time to expose thoughts and ideas and would always—or almost always—acknowledge students' viewpoints. When conversations started to drift and lose focus, she would skillfully redirect them so they would not lose track of the issues at stake.

When Cynthia had problems with specific students, she would usually not assume the students themselves were a problem. She had the habit of asking herself whether her teaching approach or the kind of relationship she had established with her students was really appropriate and investigated whether the students in question were eventually having difficulties which she was not aware of. She would also often discuss the case with some of the other teachers, and when necessary, would talk with the children's family to discuss strategies of how to deal with the problem. She did what she could to help her students in every possible way, so that the latter felt supported and successful in school.

Speaking from the "Big Self" Rather than from the "Small self"

Cynthia seemed to be able to teach from her heart, from her "big Self" (higher self) rather than from her "little self" (ego), as Jack Miller (1988) would probably put it. When she faced difficult situations with students, instead of asserting her will, as many of us probably would, she seemed to be able to step back and ask herself: "What is going on here? What is the best way to go about this situation?" Her attitude helped her maintain a good connection with students and avoid power struggles with them. Jack Miller (personal communication, 2000) explains,

> When we are teaching from the ego what happens is that we can get into these little battles with students, these little ego melodramas.... If we can teach more from the big person, then what happens is that we see ourselves in the students. The boundary between ourselves and the students breaks down. And then we can have these moments of connection with students.

Cynthia's priority was to meet her students' needs rather than winning "battles" and having things done her own way. Let me give an example of this from my field notes:

> Joshua and Daniel spent most of explore time wandering around and talking to each other. It is about 9:30 and Cynthia is about to call the group for the morning meeting when these boys bring in an electric circuit-building kit. Cynthia decides to give them a few minutes to explore the kit. After about ten minutes, she calls everyone to the meeting. The boys seem not to hear her. They keep working on the circuit-building project. She patiently waits for two more minutes, approaches them and gently calls them for the meeting once more. The boys act as if they had not heard her. She goes back to the meeting area and starts the meeting. About a minute later the boys successfully complete their project, put away the materials and quietly sit at the meeting. During sharing time they proudly talk about their circuit and

> Cynthia and some of the students ask them many questions about why the experiment didn't seem to work at first and how it came to work later on. Cynthia comments that they should probably invest more time studying electricity since they seemed to be so interested in it.

As we can see from this account, Cynthia would not impose her will contrary to students' needs. Even though in the eyes of an outsider Joshua and Daniel perhaps didn't deserve some extra "explore time," Cynthia knew that Joshua was going through difficulties and had been dealing with serious self-esteem issues. While she believed that children need strong and meaningful boundaries, she felt that at that moment, she could be flexible and give him a chance to succeed in the experiment and share his accomplishment with the group. She was open and present enough to know the boy very much needed that kind of experience.

Throughout all my observations never did I see Cynthia having students do things they did not want to do by threatening them with punishments. She seemed to always be able to convince them they had to complete certain tasks, even against their natural inclination to do so, with arguments they ended up agreeing with. Cynthia's students listened to her because, as they all told me during interviews, they liked her and thought she was fair with them. From my viewpoint, this was also the case because she usually would not "burn" herself calling their attention unnecessarily. She would only do so if she felt this would really make a difference in their learning experiences and wellbeing.

The Art and Craft of a Way of Teaching

Cynthia was sensitive in terms of perceiving what children were interested in, what their learning styles were, and what kind of challenge—and the degree of challenge—they needed in order to take their next step in terms of learning and development. She was also skillful in developing activities that integrated these different factors. It was as if she knew how to "throw a little magic" in her teaching in a way that, for the most part, students wanted and were happy to take part in her classes.

Cynthia was clear about what students should expect to happen during the day and would give them notices before transitioning to new activities, which helped her avoid conflicts with students. She was also meticulous in her observations. She would pay attention to and comment on even little things such as a poem on a T-shirt a child was wearing or small drawings children made in the corner of their essays.

Cynthia was also very good at acknowledging and valuing children's efforts. Once, for example, Diane and Clara showed her a little flute melody they had created during explore time for a school play. Cynthia said she really liked it and invited them to play the music to the group. After the girls' presentation, Cynthia tried to play the melody herself, inviting everyone to do the same. As the girls felt that their initiative in creating the melody was validated, they kept on working on it in many different opportunities until it got ready for the play.

Cynthia tended to focus on the positive, rather than on the negative aspects of children. In other words, when she was to comment on children's work or behavior, she would usually choose to talk about the good things they had been doing rather than about the negative things. She would also avoid naming the students who were not working or behaving appropriately. For example, instead of saying that some students were not doing cursive writing as she had asked them to, she would say: "I see that a lot of people are writing in cursive, and I am really excited about that...." Or, if she had been waiting for a few children to gather for a meeting, instead of calling the names of those who were not ready, she would say "so and so are ready, and we will start our meeting as soon as everyone is!" Cynthia's focus on children's qualities and on the positive aspects of their actions nurtured children's self-esteem and was healing for those who had learning difficulties and a lack of self-confidence.

The BWS is a soulful school, a special place where people want to be because of the quality of human relationships and the sense of community that permeates the whole school. It is also a place where people want to be because they feel they can be who they are and can grow according to their own possibilities. The teachers, for instance, really appreciate the school not only because they identify themselves with its educational philosophy, but also because they feel respected and valued as people and as professionals. The students, among other things, feel they can be children and explore what is important to them. The school for them is a special place not only because of the quality of the teachers' teaching, but also because of their very qualities of being.

Note

1. In fact there was one child with whom Cynthia didn't spend as much time. When I brought this fact to her, she explained she did so intentionally since this child had some learning difficulties and needed some extra time to process these difficulties on her own.

Chapter 15

The Orchestration Between Freedom and Structure

Balancing Freedom and Structure

The BWS teachers believe that children naturally want to learn and will learn as they explore things they are interested in. They also believe that, if eventually children are not invested in exploring the world around them and in learning about it, it is not because they do not want to do it, but because there is probably something else generating the problem. Based on these beliefs, they give children a large extent of freedom to explore issues of their interest. During the time frame I collected the data for my research, for instance, both throughout explore and outdoor time, they would allow children to do basically what they wanted to, as long as they were safe and did not interfere with the freedom and wellbeing of others.

But while the BWS teachers did trust children's natural interest in learning and gave them the opportunity to choose many of their activities, they tried to make sure children experienced different kinds of learning. Thus, the freedom they granted them was always framed by a well-established structure. This structure was given by a schedule of activities and by the rules that the teachers and the students constructed together. They had numerous meetings where they discussed and elaborated behavioral and work rules, which they thought were necessary, so that everyone would feel safe in the classroom and have appropriate conditions to play, work, and learn. This structure was also given by the way the teachers organized the learning environment and by the kinds of materials they made available in the classroom. The materials and activities they prepared for explore time, for example, determined to a large extent students' choices during this time: If the teachers made available only yellow, blue, and red paint, the children who decided to paint would necessarily explore combinations of these colors if they wanted to achieve specific secondary colors.

The preschool curriculum was based mostly on free play. In the case of the primary and elementary classrooms, as the children got older, the curriculum was progressively more structured, since the teachers wanted to make sure the students would work through specific skills and areas of knowledge. As I have previously explained, the primary and elementary students had a schedule of activities that included math and writer's workshops, quiet reading, and chapter-book reading on a regular basis. But even during these activities, which theoretically were the most structured of the day, they still had opportunities to make choices. For example, during reading time, they would usually decide what book they wanted to read. Or during writing time, they would often pick their own topic to write about, and rarely, they would also decide what they would focus on during their writing (e.g., the content, spelling, or grammar).

I have recently learned from Cynthia, who read a draft of this work for feedback, that as the students are getting older, their curriculum is getting more and more structured. She is now[1] working with what they call the "Elementary II" group (children from 9 to 12 years old – basically the same group of students whom she used to have and who are older now). She said that while she is maintaining a similar structure to what she had before, the curriculum is now more academic and she is a little more directive in terms of determining certain activities. She explained, for instance that now, she has implemented what she calls the "PIP's:"

> This [PIP's] stands for "personal interest projects." These projects follow a formal research process of gathering resources, notetaking, composition of a five paragraph essay, multiple forms of accompanying artwork, and a formal schoolwide public presentation where each student reads his or her report to the school community. (This makes for heightened purpose and accountability in their writing.) This is called a "PIP Event." We do a "PIP and PIP Event" in the fall. We do a "Science Fair PIP" in the winter, and also a spring "PIP and PIP Event."

Cynthia also explained that now that her students are older, she is a little more directive even during explore time. She often requires that they finish up incomplete work from the previous day or do some other quiet assignment that she lays out at their desks before they arrive (e.g., individualized math practices and/or some kind of spelling games). Only after students complete their tasks are they truly free to make their own choices of activities. According to her, some of their favorite choices have been art activities with open-ended materials, play rehearsals, creative writing or journal writing, knitting little crafts to be sold at the Solstice Fair to raise money for the local food shelf, and the development of the PIP's.

Issues of Accountability and Evaluation

As the BWS curriculum was based on subjects children were interested in, a question that permeated my data collection was: Did the teachers make sure to cover the kinds of knowledge and skills children were supposed to develop according to their developmental level, and if so, how did they do so? According to my interviews and observations, both the primary and elementary classroom teachers consulted official standards and grade books to keep them informed, help them set goals, and orient some of their practice. They also partially adopted a math program which gave them suggestions of activities to develop with students according to their level of development.

Another question that permeated my data collection was: "How did the teachers keep track of students' work and evaluate them considering that individual students were often involved in different projects?" For the elementary classroom teacher, keeping track of students' work and evaluating them did not seem to be too hard a task since she was the only teacher in the classroom and had only nine students. In the beginning of quiet reading and of the writer's workshops, for example, she would always have each of them tell her what they would be working on during these specific times. She would then go around and work with individual children or with small groups of students according to the latter's developmental level, interests, and needs. At the end of the reading section, she would have students fill out a book record (a piece of paper where they listed the names of the books they had read), and from time to time, she would also have them write a "book report" (an essay about the book they had finished reading).

It was harder for the primary classroom teachers to keep track of students' work than it was for the elementary classroom teacher. Considering that they were two teachers with 18 students in the same room, they did not always seem to be able to know what exactly each child had done by the end of the day. It was also hard for them to know whether the other teacher had worked—and on what exactly he/she had worked—with each individual student. Perhaps, a helpful strategy for these teachers to take care of this situation would be to keep a written record with the names of all the children, where they could systematically note down the dates when they had worked with each child, what exactly they had worked on, and what the child was expected to do by the next time they met again. This way both teachers would always know what kind of attention each child had had, and how they could best follow up with the work of the other teacher.

The BWS students did not take tests. Evaluation was done on a continuous basis through the teachers' observations of each child's learning and development, which was officially reported to parents both in written form and in individualized conferences with them. Cynthia, for example, had two one-hour individual conferences with the parents of each of her students, one in October, and one in March. She would also individually meet with them whenever necessary to discuss specific issues. At the end of the year, she completed a checklist and wrote a descriptive evaluation of each child addressing his/her learning and development in all areas of the curriculum.

An interesting aspect of the teachers' pedagogy and evaluation was that they consistently documented everything that happened in the classroom with pictures. The pictures were, for them, a "living history of the classroom." They helped them keep track of their work and students' progress, rethink actions, and plan future activities. They also facilitated conversations with the students about their work as well as helped them keep the school community informed about what had been happening in the classrooms (each of the classrooms periodically organized murals with the pictures).

Students', Parents', and Teachers' Perspectives on Freedom and Structure

The BWS students seemed to really appreciate the kind of balance between freedom and structure that was provided to them at the school. I interviewed all the students in Cynthia's classroom on this topic, and all of them said they really liked the way the teachers structured their schedule and activities. See for instance the following comment by Diane (9 years old):

> I like how it is all put into a very structured curriculum but it is also sort of smooth going and flexible in the classes. So like there is a curriculum that is strict and it says "we have writer's workshop on Tuesdays and Thursdays and math on Mondays, Wednesdays, and Fridays and music on Wednesdays." But then between these activities it is really flexible, ... lots of different activities.... I like how it [the day] is put together. It runs out very smoothly!

When I asked the elementary classroom students whether they would change anything in the structure of their schedule and activities in school, eight out of nine students said they would not change anything, which indicates that they really liked the school exactly the way it was. Only one student said that she would perhaps prefer to have "quiet reading" first thing in the morning rather than "explore time," because she tends to be a little "slow" at this time of the day. It is interesting to note, however, that if this child really wanted to read early in the morning, she was free to do so during explore time. However this was never her choice. She would rather choose other activities.

I interviewed seven parents (the parents of seven children from Cynthia's classroom). Six of these parents said they were completely satisfied with the balance between freedom and structure in the school, and felt that it worked quite well for their children. Only one parent said she wasn't sure the school was the right place for her child. She explains:

> Joshua was starting to fail and to shut himself down in the public school. At BWS, he regained a sense that he does have strengths and that they are important and appreciated. This helped his self-esteem and self-confidence that were quite low when he started in the school.... But now he is becoming keenly aware that he can't produce like other people and he doesn't want to feel odd. So I don't know ... I don't know how to get a child to produce.... But he certainly needs to have a more consistent requirement.

While this mother was happy for what the school had done for her son in terms of helping him regain his self-esteem and self-confidence, she also felt that he probably needed more structure than that provided by the BWS. She thought that he perhaps demanded stronger directions in order to work harder and overcome some of his difficulties. Consequently, she ended up moving him to the public school system in the next academic year.

I intensively discussed with Ron Miller (who at the time was the head of the Board of Directors) and with Toby (one of the primary classroom teachers) whether they thought the structure of the BWS addressed the needs of every child. They both stated that it didn't, since many children demand more one-on-one attention and/or more direction in terms of what to do in school and how to work than the BWS offered them. This was in fact the case of Ron Miller's eldest child. As I have previously stated, he helped start the BWS elementary program motivated by the idea of enrolling his son in the program. Ironically, however, he found that this educational model proved to be too unstructured for his child's learning style, and by second grade decided to transfer him to a Waldorf school. Even though Miller had some reservations about Waldorf education, particularly its highly structured environment (which tends to guide children step by step in everything they do at school), he reached the conclusion that the degree of structure this kind of education offered was necessary to help his own child optimize his learning and development.

As Cynthia read my work and gave me feedback, she said she fully agrees with Toby and Ron when they say that the BWS structure does not meet the needs of every child. She also added to this viewpoint the interesting idea that, for a school to work, it should not only meet the needs of the child but also satisfy those of his/her parents. If the latter keep thinking about fitting the child into a standardized system rather than to continue on with alternative education, chances are that the child will have difficulties in school.

I interviewed all the BWS teachers about their beliefs on freedom and structure in education and found they had different views on the topic. Andy—one of the primary classroom teachers—for example, tended to be a little looser in terms of structuring the schedule and activities than the other teacher (Toby). He would naturally prefer to be in the moment with students rather than follow the structure of the schedule of activities. For instance, if he were playing an interesting game with the students, he would usually want to continue the game rather than calling the group for the activity planned to happen. While Toby was flexible and did adapt the schedule to the emerging needs of the moment, she tended to be sharper in terms of making sure children would have their math, reading, and writing practices on a regular basis. While I was doing my research, the differences between these teachers didn't seem to be a problem since Andy tried to adapt to Toby's style (he was a new teacher and felt he had a lot to learn with Toby who had been teaching for many years). I wonder however, if the differences between them will not lead to problems in the future to the extent that the students start noticing the inconsistencies between their practices.

Before ending this section it would be interesting to mention here that when I asked several parents what they liked the most at the BWS, one of the things they all mentioned was that it was nurturing in their children a love for learning. One mother commented,

> The gift that I feel [the school is giving to my child] is this incredible love for learning.... There is no question about coming to school, there is no question about doing homework. There is no question about

> reading or developing projects. And those are always first choices versus other kinds of play.... She doesn't watch television.... She might write her own song, or make up a song in the piano, or make up a game. So I like what I see there, versus do I know whether she is at the right reading level or writing level. Or is she getting all the history that she needs at this point? I like the way she lives, I think it is wonderful, and I think that if she builds that baseline, when she is challenged, I think she is going to be confident enough to learn it.

I asked the parents whether they were concerned with the fact that, given the structure of the BWS curriculum, their children perhaps were not covering as much in terms of content and subject matters as students from other schools. They all replied that while they sometimes felt a little concerned about that, for them it was more important that their children could still enjoy their childhood while at the school, that they could explore issues they were interested in, that they were learning how to learn, and that they were developing a love for learning. This, they said, would allow them, in the long run, to become healthy, independent and passionate learners.

Note

1. A year and a half after I collected the research data on the BWS.

Chapter 16

Final Comments

The Multi-Age Aspect of the BWS

The multi-age and non-graded aspect of the BWS curriculum was beneficial to children in many ways. As a flier of the BWS explains,

> In non-graded classrooms, there are less rigid timetables for student progress, so competition is reduced and students of various abilities are free to interact as unique individuals; they help each other and learn from each other. Students have time to explore their own interests, at their own pace, according to their own learning style. Because teachers are with the same children for more than a year at a time, they are more attuned to each child's long-term development, and children do not experience disruptive transitions every September....

Another advantage of the multi-graded classrooms of the BWS was that students learned to live with and respect the gifts and limitations of peers of different ages. The younger children learned a lot from the older, and the older learned by teaching and helping the younger. Another advantage was that, because in that environment there was an implicit assumption that everyone was at a different developmental level, little room was left for comparisons among children. Therefore the latter were less likely to be singled out because of their differences and difficulties. This helped them feel more at peace with who they were as individuals. During an interview Cynthia actually talked about this by giving the example of a specific student:

> I think of a child who had a very severe processing disability and was tested and diagnosed as such, and wanted very very much to learn to read. And yet this was enormously difficult for him.... At this school, because children are at different ages and weren't standardized in a way that said 'at this point you have to be doing this or you are a failure ... at this point you've got to be doing that, or you are a failure" ... now this child is an avid and accomplished reader. ... I think that because he wanted it so much, because he was in an environment that wasn't in any way threatening, and all the way a sense of success was there wherever he was—no teasing, no comparison with other children—all of those things allowed for that success.

From my perspective, the multi-age classrooms of the BWS worked well for two main reasons: First, the low student/teacher ratio, which allowed the teachers to work on an individual basis with children. Second, the degree of commitment of the teachers to promote a sense of collaboration between the students. It is interesting to note however, that the multi-age environment of the BWS did present one drawback: Sometimes the younger students, especially those from the pre-school, had to sit through long meetings with their older peers. They sometimes seemed tired and also a little overwhelmed with the amount and complexity of issues discussed throughout these meetings. Perhaps a good way to address this problem would be to separate the preschool in to two groups during these meetings—one with the younger and one with the older students—so that they could be better adapted to the developmental needs of the different ages of students.

Transition to Other Schools

Because the BWS is relatively new, there is little data on how students who left it are doing in other schools. Nevertheless, I did talk with a few parents who told me that according to their experience—or to the experience of people they knew, the BWS students who had transitioned to public schools were doing well. Only one parent said that her child had had a different experience. According to this parent, her son felt bored in the public school since he was more advanced than his peers in many ways and the teacher "held him back" by not attending to his differences. According to her, the child also resented the "transmission model" of the public school where he had too few opportunities to explore issues he was interested in and build his own knowledge, as he was used to at the BWS. As he did not seem to adapt in the public school system, she enrolled him back at the BWS.

An interesting lesson to be learned with the BWS experience is that a focus on the development of the child as a whole being and in all her aspects does not necessarily take away from child's intellectual development. It is true that perhaps the BWS curriculum does not cover as many different subjects as other kinds of schools might cover. However, because the BWS students work with subjects they are interested in, and because the teachers provide them with a structure that leads them to explore in depth these subjects, they not only progress in terms of knowledge and abilities, but also tend to develop a love for learning. These knowledge, abilities, and especially love for learning give them, in the long run, conditions to become independent life-long learners.

The Issue of Freedom and Structure

As the BWS teachers believed in children's natural interest in learning, and believed as well that they would learn if given the chance to explore things they were interested in, they gave them a large extent of freedom to work around projects that were attractive to them. The freedom they gave their students was, however, framed by a well-established structure in order to make sure children would be exposed to different learning experiences; and in the case of the primary and elementary classroom students, that they would cover the different subject areas and practice the basic academic skills (reading, writing, and math). The structure that determined students' experiences was given by a schedule of activities, by rules constructed by the teachers and students, by the way the teachers organized the learning environment, and by the materials they made available in the classroom.

The preschool children could select their own activities throughout most of the day providing they were safe and not interfering with the freedom and well-being of other students. During the time frame of my data collection, the primary and elementary classroom students could select their own activities during about one-third of the day (explore time, outdoor time, and lunch time). During the rest of the day they were guided through specific activities so that they would cover different academic skills and areas of knowledge. The oldest group of students—the "elementary II group"— which did not exist at the time of my research, according to Cynthia who has recently written to me, also has a similar structure to their day. However, they are a little more directed in terms of the kinds of activities they have to necessarily engage in on a daily basis. But regardless of the level of guidance they are subjected to, they still have numerous choices in terms of how to go about accomplishing set goals.

Before closing up this section, it is important to remark, as Cynthia once commented after she read a draft of this work, while the BWS is guided by some values, it remains eclectic and open, and therefore, subject to changes. Therefore, "any glimpse of the Bellwether School is a glimpse of a work in progress." The structure I observed at the time of my research might be different as I am making this observation now.

THE FUTURE OF THE BWS

According to my observations, the BWS is indeed a wonderful school. Nevertheless, as a researcher I do have a major concern for its future. As I have previously stated, the school is relatively new and still doesn't have a solid educational philosophy and a clear position on the issue of freedom and structure. This, plus the fact that the group of people who run the school is changing over time, may lead the school to, in the long run, lose track of the vision that has led to its creation.

Actually the school seems to have already significantly changed after I finished the data collection. A few months after I left the school, I had the opportunity to meet with one of the teachers who told me about some innovations. Some of them, for example, were that the curriculum would include classes with specialist teachers for Spanish, arts, and sports (the students would rotate in swimming, tennis and soccer classes). While these were changes that teachers, students, and parents were looking forward to, I wonder to what extent these classes will take away from children's time to explore issues of their interest, as well as chop the schedule and fragment the school experience as it has been happening in most traditional schools.

Chapter 17

Closing Remarks

So far, I have presented independent case studies of each of the schools included in my research. In this final chapter, I will compare the three schools and deal with a question, which I kept asking myself as I developed my study; That is, which school, from my viewpoint, seems to best address the interests and needs of children? In the end, I will make some recommendations for future research in the field of holistic education.

The main similarities among the SAU, the SWS, and the BWS:

- They fostered the development of the whole child. In other words, unlike most traditional schools, they did not prioritize the cognitive/intellectual aspects of children over other aspects (physical, emotional, social, and spiritual).
- They encouraged students to explore different abilities, potentials, and a large repertoire of competencies.
- They nurtured a sense of connection between the teachers and the children, between the children and what they were learning, and between the children and the natural environment.
- They promoted children's appreciation for the beauty of the world around them.
- They protected childhood by letting children be children, by letting them play and enjoy their experiences in school, and by not overloading them with schoolwork or homework.

- They had a well-flowing schedule, where children could usually explore the activities they were involved in without having to hurry from class to class.
- The different subject matters were taught within an interdisciplinary approach.
- Teaching and learning were based on hands-on experiences rather than on the mere transmission of knowledge.
- The teachers focused more on the qualitative aspects of students' learning (how they were learning) than on the quantitative aspects (how much they were learning).
- The assessment of children's learning and development was continuous and organic: It happened on a regular basis as the teachers observed/interacted with the students.
- They acknowledged children's need for physical movement. Both the SAU and the BWS students had outdoor playtime for a minimum of 45 minutes a day. They could freely move around in the classroom as long as they were doing their work and were not negatively interfering with the work of their peers. While the SWS students had less outdoor playtime and less freedom to move around, their curriculum encompassed activities throughout the day that promoted physical action (e.g., the morning circle, the dramatization of stories, the art classes, and the "singing with movement" that went on during various parts of the school day).
- The teachers stayed with the same group of students for a number of years. In the case of the SAU and the BWS teachers, they taught the same group of students for two, three, or more years. In the case of the SWS teachers, they usually worked with the same group of students for eight years.
- They were deeply committed to fostering a sense of community between all those involved with the education of children (teachers, students, and families).
- They promoted a sense of collaboration between students. This, plus the fact that they were located in states that did not mandate that children in private schools be tested or graded, helped them

keep an environment where children did not compete for grades and were not compared with each other in terms of performance.

- The faculty had autonomy to develop the curriculum according to the rhythm of their classroom (they did not have to follow a weekly or monthly schedule of activities).
- The teachers played a strong role in the administration of the school.
- The parents of each of the schools tended to have a sense of shared values with other parents and with the school staff in general. This led them to deep friendships with other people involved with the school as well as motivated them to volunteer in school committees, classroom activities, and in events such as field trips, plays, and festivals.

The Main Differences Between the Schools

Curriculum and Spirituality

According to Ron Miller (1993), one of the main characteristics of holistic education—what distinguishes it from other approaches to education—is that it addresses the spiritual dimension of human development. While the three schools where I did my research did take this dimension into consideration, they did so in different ways and degrees:

- The Sunshine Waldorf School—as Waldorf schools in general—had a well defined ontology and educational philosophy. Waldorf advocates believe that we are spiritual beings who come to life in a series of reincarnations in search of self-development. Therefore, Waldorf teachers see their role as helping children in this journey, which they try to do in different ways. For example, they supposedly work on their own self-development on a regular basis, hoping not only to improve themselves but also to become good role models for their students. Besides, they implement a curriculum that includes, varying with children's age, fairy tales, stories of saints, mythology, biblical stories, and the study of the principles of different kinds of religions. These subjects, according to them, lead children to directly or metaphorically address issues of a spiritual nature. Also, the morning and the end of the day verses, the prayers before snack and lunch, and all the songs they sing that talk about a higher power that exists beyond ourselves potentially foster in children a sense of awareness and reverence for life, and nurture in them a feeling that everyone and everything in the universe is interconnected and a part of a larger and transcendental whole.

- Differently from the SWS, the expression of spirituality in the SAU and at the BWS was less direct. Both the SAU and the BWS teachers did not openly talk about spiritual matters as part of the curriculum they offered children, and when asked about it, they tended to express different understandings on this issue. Most of them admitted not to have a clear view of what exactly spirituality in education meant for them and how it was expressed in their practice. But regardless of their lack of clarity on this matter, they did nurture a sense of spirituality in their classroom in various ways. For example, like the Waldorf teachers, they fostered meaningful connections with children and between children, between children and their inner-self, between children and the natural environment, and between children and what they were learning. They also respected the individuality and uniqueness of each of their students and fostered authenticity and creativity in the classroom. And also very importantly, some of them modeled powerful and spiritual ways of being in the world as they expressed their love for children and for their work, cultivated a mindful presence in their relationships, and showed an appreciation for the beauty of the world around them.
- An interesting characteristic that distinguished both the SAU and the BWS curriculum from that of the SWS, was that while their focus was on the local and on the moment, the focus of the SWS was on the transcendental (spiritual reality). In other words, the SAU and the BWS teachers focused on the "here and now" of what was going on in children's lives and in the world around them. They believed that it was important to keep children

informed of current events, address in class the issues that children said they were interested in, and encourage them to make decisions about and take action on important matters that pertained to their lives. The Waldorf School teachers, on the other hand, acted as if children needed to be protected from the "here and now." They believed this kind of protection was necessary, if they were to be thoroughly prepared to deal with their own lives and with the world around them at some point in the future in ways that would help them become better than they already were and that would help them transcend their current reality.

- Another interesting characteristic that distinguished the three schools was that while the Waldorf School openly criticized the materialistic mentality and consumerist values that permeate our society and called for a radical transformation of the world we live in, both the SAU and the BWS didn't directly address this issue. The SAU and the BWS staff had different views on this matter and didn't explicitly deal with it in their educational philosophy and practice. Apparently it was easier for the Waldorf teachers to have a clearer standpoint on issues such as this one, since Waldorf Education has been around for more than 80 years, and all Waldorf teachers go through similar training programs and follow the orientations of one, and only one, major leader, Rudolf Steiner. Both the SAU and BWS are newer schools and their teachers go through different kinds of training programs, study different schools of thought, and do not have a central author or a director/leader/administrator guiding their work. This generates an eclecticism in their thinking and practices.

Pedagogical Aspects

- While the SAU and the BWS curriculum was emergent and everything children were interested in could easily become a part of it, the SWS curriculum was almost completely planned in advance and there was little room to address subjects that eventually emerged in class. A good example of these approaches could be seen in the ways in which these schools treated my presence in the classroom. The SAU and BWS teachers saw my presence as a possible learning opportunity for them and for their students, and therefore, often invited me to take part in different activities. The SWS teachers, on the other hand, whereas they did welcome me into the school, they saw my presence as potentially disruptive. They would usually have me sit in a corner of their classroom and would ask me to interact with students as little as possible so as not to alter the natural unfolding of their work.

- While the teaching both in the SAU and in the BWS was either individualized or in small groups (the teachers worked according to the needs and developmental level of each child and/or of each small group of children), the teaching at the Waldorf school was usually directed to the whole group of students. Individual children had to adapt to the level and rhythm of work of the majority of their peers.
- Both the SAU and the BWS teachers and students had a close relationship (they acted more as friends than teacher and students in the traditional sense). For example, the children would call them by their first name and they would often have conversations about personal issues (things that were going on in their lives outside school). They also usually had a significant amount of physical contact. They sat close to each other in desks, tables and/or floor and not rarely would hug and/or hold hands as they talked. The Waldorf teachers, on the other hand, had a more formal relationship with the students. The children referred to them by their last name and teachers and students usually did not talk about personal issues in the classroom. Besides, they also kept a certain distance in their sitting arrangements. Whereas the teachers sat on a special desk in front of the class, the students sat in little desks disposed in rows carefully distributed by their teachers along the classroom. The Waldorf teachers purposefully created and maintained a picture of themselves as authority figures for children so that they could fully implement the Waldorf philosophy and direct the latter's activities as much as possible in school.

Thinking and Feeling

- Both the SAU and the BWS teachers aimed at fostering students' thinking and intellectual skills. They would, for example, ask children questions such as: "What do you think about this?" "How did you come to this conclusion?" "Do you think this is OK?" "Why?" The Waldorf elementary teachers, on the other hand, tried as much as they could to protect their younger students from thinking about and from intellectualizing their experiences. According to them, children should first develop their "will" and "feeling life, " from 0 to 7, and from 7 to 14 years old, consecutively. They believe that the focus on children's intellectual development should not start before 14 years old. A good example of the differences between the schools' approach on this matter can be seen in the following quotes which I extracted, one from an interview with an BWS teacher and one from an interview with a SWS teacher. The question I posed to them was the following: "What would you do if a child asked you 'why is it raining?'"

If they are asking me why is it raining or where is the rain coming from? To me that's great! And I might even ask that question back to the group.... Let's hear all of the different ideas.... This is really that's what the sciences are all about, asking questions.... If they are expressing that maybe there is a God up there just dropping rain down or whatever their theories are, they are still valid and important.... I might talk about research and what scientists do. I think I would probably bring it to a more scientific level of how people find these things out, if the kids are really interested. I might go measure the rain and see how much there is, if they are deep into it. There is the depth of rain, what kinds of bugs come out and all that kind of thing.... There are so many places you can go with that question.... (Sara, BWS preschool teacher)

I generally get more "why's" in the beginning of the year than later, because I don't answer them all....We don't want them to get into their heads too much when they are in early childhood and start to think

> and reason too early. If they are going to stay in the world of more dreaminess and imagination, then you don't want to put them down that early with logical figuring things out.... So if the child asks me why it is raining there are two ways I might go, and one way might be just to say: 'Hum, why do you think it is raining?' And just let them ponder.... Sometimes I might just give a really simple 'oh, because that's what they are doing today.' In fact most of the time that is absolutely fine with them. Sometimes the 'why questions' are coming because they want attention. They just want that time with you. It is not that they really have some burning need to know all of a sudden. We also may create imaginative pictures and so then the children can live in the imagination the elemental beings, fairies and so on.... (Mary Lee, SWS nursery teacher)

As we can see in the examples above, while Waldorf teachers working with pre-schoolers—and I would also say, working with elementary school children—would want to avoid working with children on the scientific concept of rain, the BWS teachers—even the pre-school teachers—would fully explore this concept, if children showed a strong interest in it. In the beginning I didn't understand the Waldorf perspective and tended to criticize it. After all, why not stimulate children's scientific thinking? Only after I had an experience with my own child that spoke to this matter, did I come to understand the Waldorf approach. When my daughter was to turn four, one of the things that she liked the most was to watch the train go by the window of her grandmother's apartment. As soon as she would hear the train coming down the rails, she would run to the window and stare at it full of joy and wonderment. One day I instinctively started counting the wagons aloud as we watched the train. Since that day, unfortunately, every time she sees the train, she insists that I count the wagons with her. I said "unfortunately" because, since then, as she focuses on the counting, she seems to miss some important qualities of the "train experience" that she used to be more attuned to. She seems to no longer care about the sounds of the train, the colors of the different wagons, the vibration of the train on the ground, the good-bye waving of the train engineer who sometimes notices us at the window. She also doesn't seem to mind any more about all the wonderful stories we used to create on what could possibly be inside the wagons.

This experience taught me an invaluable lesson that goes along the Waldorf perspective: The premature intellectualization of children may be detrimental. Children should have the opportunity to be in wonder with and "feel" the world around them in all its plenitude before being asked to intellectualize about it. The intellectual development of children is obviously important and should be stimulated. However, it seems that we should avoid hurrying children towards this goal so they will not miss other important aspects of development, which perhaps, they will not have as much opportunities to develop later in their lives.

Impression and Expression

- Both the SAU and the BWS teachers encouraged students to express themselves. They had children talk about things that were going on in their lives and explore subjects they were apparently interested in. The Waldorf teachers, on the other hand, instead of having children express themselves, they tried to "impress" in them a certain worldview and way of being. According to these educators, much of what children want to express is imposed on them by the media and different people in their lives. Therefore, they believe students should first "drink" from sources of "wisdom," "rightfulness," "truthfulness," and "wellness," and only later, as they became critical of certain "influences," should they be encouraged to express themselves. A good example of the differences between the SAU and the BWS when compared to the SWS could be well observed in the ways in which they approached the teaching of the different arts. While the SAU and the BWS encouraged children to fully express themselves through different artistic means, the Waldorf teachers would tell their students exactly what to do and how to do their art work in minimal details (e.g., how to hold the brush, what colors to use, in what directions they should move the brush on the paper). They believed that only as children learned to use certain tools and techniques, developed a taste for beauty, and developed a true vision of reality, would they be able to fully and freely express themselves.

Freedom and Structure

- The understanding of freedom both at the SAU and at the BWS was very much related to the concepts of democracy and choice. But while at the SAU the students had autonomy to make significant decisions about their school experiences, at the BWS, the students made choices only within the framework given by their teachers, who tried to make sure the children would regularly and systematically work through specific knowledge and skills. Differently from the other two schools, at the SWS the understanding of freedom was not related to the concepts of choice and democracy. Waldorf educators are of the opinion that freedom has to do with an inner state of mind, with a sense of groundedness on one's innermost being, and with a critical standpoint in relationship to hegemonic ideas, goals, and materialistic concerns. They are also of the opinion that freedom involves self-control, self-discipline and a balance between the faculties of will, feeling, and thinking. According to them, it is only through close guidance that children will develop these qualities and will one day find their own inner-teacher, that who will ultimately lead them to utmost freedom.
- While the SAU and the BWS teachers encouraged students to make numerous decisions about their school experiences, the SWS teachers guided their students step by step in all their activities. The latter believed that only as children grow older and more mature, and only as they show an ability to make authentic and wise decisions on their own, should they be allowed to make certain decisions about their school experiences. A good example of how the schools approached this matter differently could be seen in how while both the SAU and BWS teachers welcomed my idea to write a letter to students informing them of my research and asking for their formal permission to observe and interview them, the Waldorf teachers asked me not to do so. From the latter's perspective, elementary

school children are not prepared to make this kind of decision, and therefore, should not be asked to do so.

My Personal Stand on the Issue Of Freedom and Structure

I would like to go back to the authors I mentioned in the introduction of this book and relate their work to my own view on freedom and structure in education. From Krishnamurti (1981, 1995) Steiner (1920/1995, 1998) and others—specifically the eastern philosophical tradition—I take the notion of "inner-freedom": the idea of freedom as self-knowledge, self-awareness, and the absence of conditionings. From Steiner I also take his emphasis on strength of will to do what needs to be done as key for a thorough understanding of freedom. I believe that it is very unfortunate that these views of inner-freedom are not emphasized in the Western world. From Dewey (1964) I take the idea of freedom as related not only to choice and objective conditions, but also to intelligence and knowledge, since the latter determine our ability to thoroughly observe, to foresee the consequences of our choices and actions, to make plans and judgements, to be effective in what we do, and to change the course of our plans and actions as needed. From A.S. Neill (1960, 1966), Greenberg & Sadofsky (1992), and Holt (1972, 1982, 1983), I take questions which I keep asking myself such as, "What right do we have to tell children what and how they should learn?" and "How can we say that what he [she] wants to know is less important than what we want him [her] to know?" But while I keep asking myself these questions and I do think schools like the School Around Us work for certain children, I personally believe that as a general rule, most children do need a degree of guidance in their educational experiences so that they can be truly free—both inwardly and outwardly—for various reasons:

- No matter how much we hope children can think, act, and make choices for themselves, they will always be influenced by other sources of "guidance." Their peers, television, videos, Internet sites, and the media in general will inevitably lead them in directions that aren't always beneficial. As Jeffrey Kane once said (personal communication, 2000), without some kind of clear standpoint and direction by the part of schools,

> It is possible, for example, that kids would be interested in what the media feeds them. It is likely that they will be interested in what merchandising and corporate America designs to sell them. It is likely that schools will teach a very structured set of values like individualism, materialism, consumerism, reductionism, etc., even if children might think that they are free from structure in following their own desires. *There is always going to be a structure in schools. The only questions are who is going to set it up, on what basis and to what end.* (emphasis mine)

- Many children do not have a good idea of what they are capable of, and if not given a hand, they may never transcend their views on some of their potentials and current reality (this is especially true for children who have low self-esteem). A degree of structure is important so that they can go beyond simple self-satisfaction and reach self-transcendence.
- Whereas for some children the school experience might be the only instance where they can explore their humanity and unique selves, for many others schools might be the only place where they can have the opportunity to develop specific skills and have access to certain bodies of knowledge. This is especially true for children who live in poverty and/or live in homes that do not offer them enough positive learning opportunities. (Saviani, 1984; Mello, 1982)
- A degree of structure is important so that we can make sure children will engage in a variety of rich learning experiences. As Dewey (1963) contends, what children like to do and choose to do is influenced by their previous experiences and by what they have already learned. The richer the learning experiences a child has been exposed to and the more she knows, the more likely she will be to want to involve herself in new and more complex learning experiences.
- A degree of guidance is important so that we can develop focus, self-discipline, and strength of will to achieve long-term goals such as the learning of reading, writing, and research skills. These skills, as we know, determine our possibilities to have certain options—and to make concrete choices—about our

professional life (outward freedom). A degree of guidance is also important so that we can develop regular and systematic practices (e.g., purposeful reflection, contemplation, and meditation). These practices, as eastern traditions teach us, can help us develop self-knowledge, self-awareness, and self-control, which in turn can free ourselves from social, cultural, and internal fears and conditioning (inward freedom).

- We need to make sure all aspects of the child—intellectual, physical, spiritual, emotional, and social—are honored and addressed. Otherwise we might unconsciously focus on those which are most appealing to us, leaving other important aspects aside.[1]
- Last but not least, children need structure and guidance to help them learn to limit their personal demands so they can sustain healthy relationships with other people and with the natural environment, and so they can learn to live with limits if we are to ensure ecological (e.g., Bowers, 1993; Orr, 1994; Smith, 1992) and cultural sustainability (R. Miller, 1995).

From my perspective, independently of our position on the issue of freedom and structure, if we are to start a school and/or already are part of a school community, we need to make it public what our position on this matter is. This would be important so that teachers and families who are planing to get associated with the school can have a clear view of what the school is about and whether this is something they really want for them and/or for their children. This would also be essential so that the work of each teacher can be consistent with the work of the other teachers and so that the school can avoid the kinds of problems described in the first section of this book.

Some General Comments and Recommendations

- It seems that both the SWS and the BWS would benefit from introducing a regular community service program at the elementary school level (at the time of my research, they were involved only in some seasonal service programs). Community service can serve both as a springboard for academic learning and also as a means to help students develop a sense of social responsibility (for more on community service learning see Coles, 1993; Dwight, Honnet & Migliore, 1991; Hedin, 1989; O'Neil, 1993; Swanson, 1994; Stanton, 1990; and Silcox, 1993). As we know, many children today—especially teenagers—do not know what to do with their free time besides staring at a TV set or hanging out in malls. Getting involved in social service could possibly help them develop a sense that they can make a difference in other people's lives, which in turn can bring some meaning to their existence.

- Some of the SAU and of the BWS teachers had a few confined animals in their classrooms. These teachers believe this can nurture in children knowledge about and a sense of caring for animals. While this might in fact happen, this practice is cruel to the animals that are kept confined, despite teachers' good intentions. This unconsciously sends children the message that it is OK to take the freedom of other living beings away just to satisfy our needs, which in turn contributes to the growth of an anthropocentric worldview, a view that leads us think that everything in the world is here to satisfy our needs.
- We as teachers usually make little money when compared with other professionals with the same level of education in different areas. We end up subsidizing other people's education with our low wages. Unfortunately this reality was also present in the three schools where I did my research. This needs to change so that we can make a more decent living and have more time, energy, and resources to dedicate ourselves to our personal and professional development, which will inevitably revert in benefits not only to us, but to our students as well.

Which School is Best?

In face of the differences, gifts and limitations of each of the schools discussed in this work, at this point we may ask ourselves, but which of the three schools is best? From my perspective, there isn't one, and only one right answer to this question. In order to answer it, we need to take into consideration at least three main variables: (1) What kinds of values do we hold as a family and what kinds of expectations do we have for our child? (2) Who is the child and what kind of structure does he/she need in order to optimize his/her learning and development? and (3) Who is the teacher and how does he/she relate to students?

What Kinds of Values Do We Hold as a Family and What Kind of Expectations Do We Have for Our Child?

The SAU, if it were not for the problems that were happening during the time of my research, could be a good option for families who deeply believe in the kind of education fostered at the school and live this kind of education at home. As I observed in the SAU—I actually observed this in the SWS and in the BWS as well—the children who came from families who fully lived the school's philosophy at home tended to feel more comfortable and to do better in school than the other students. Those who attended the school only because the school was convenient in terms of price and/or location, or because they couldn't adapt in any other kind of school, didn't seem very comfortable and/or didn't do well academically and/or behaviorally.

The BWS and the SWS would be good choices for families who believe in the educational philosophy of these schools—and live them at home—and want to make sure their children will necessarily work through certain academic skills by the end of specific cycles of schooling. The BWS would probably be a better option for families who believe that children should have a voice in their education and actively participate in decision-making processes.

Waldorf education would probably be a good choice for families who: (1) believe that a sense of spirituality should permeate everything children do at school; (2) would like to openly reinforce values such as anti-materialism and anti-consumerism; (3) want their children to experience, on a regular basis, not only a rich academic curriculum, but also the study different religions, foreign languages, musical instruments, eurythmy, handwork, woodwork, singing, drawing, painting, and modeling with clay. Waldorf education, however, would probably not be a good fit for families who would like their children, from an early age, to be informed of current issues, become versed in technological skills, and practice participating in democratic processes. As I have previously explained, Waldorf schools usually do not give their students much space to participate in decision-making processes in the classroom and protect them from information and situations that they consider "overwhelming" to children. They also do not expose their students to certain technological tools until they feel the latter are fully prepared to deal with them (e.g., they will usually not allow the use of computers before high school).

It is important to remark here that, independently from what kind of school we choose for our child, if our choice is to be a good choice, we need to be committed to its educational philosophy. When there are doubts about the kind of work developed by the school, and when there are inconsistencies between school practices and home life, there is a good chance that the school will not work well for the child.

Who is the Child and How Does He/She Learn Best?

Some children need more structure than others to optimize their learning and development. Waldorf schools would probably be a better option for children who need a lot of structure in order to learn. Both the SAU (if it were not going through the problems described in the first section of the book) and the BWS would probably be better choices for students who tend to be more self-directed and independent learners The experience of three siblings who I met during my research can well illustrate this point.

The three brothers used to go to the BWS. However, in time, both their teachers and parents noticed that the oldest child, George, didn't seem to be doing well in that environment: He wasn't able to focus for very long whenever his teachers were not directly working with him. As they realized that George clearly needed more structure than the BWS could provide him, his parents moved him to a local Waldorf school.[2] They found that this school was a better choice for him since there, differently from the BWS, he was guided step by step in all school activities, which was exactly what he seemed to need in order to focus and do his school work.

But while the Waldorf structure worked well for George, it didn't work for his younger brothers, Robert and David. Their mother explains:

> My two younger children have a very clear set of ideas of where they want to go with something and need the freedom to explore those ideas. That if they want to write or do work around King Arthur for instance, being told the exact words that they need to write down on a page didn't work for them because they need to be able to say, what is it for me that King Arthur represents?

The Waldorf structure in fact felt "suffocating" for Robert and David , who are independent and self-directed learners. As the Waldorf experience seemed not to work for them, their parents moved them back to the BWS, where they are still studying and doing quite well.

The experience of the three siblings suggests that we as teachers and parents need to be fully attuned to the fact that some schools work better than others for different kinds of learners. While most of us could not afford to have our children in different kinds of schools as the family just mentioned could, we do need to consider their learning styles and the kind of structure they need to optimize their learning and development so that we can make well informed choices for them.

Who is the Teacher and What Kind of Relationship is He/She Able to Establish With the Child?

Many authors, including Scott Forbes (personal communication, 2000), R. Kessler (2000), Krishnamurti (1981), Levy (1996), Ba & Josette Luvmor (personal communication, 2000), J. Miller (1993, 1994), R. Miller (1995), Palmer (1993, 1998), and Steiner (1998), among others, have noted the importance of the "quality of being" of teachers. In other words, they have remarked that who we are as a person and how we behave around children have some kind of impact on students. See, for instance, the following quotes:

> These are the things that matter the most for young children. What you say, what you teach, does not yet make an impression, insofar as children imitate what you say in their own speech. But it is what you *are* that matters; if you are good this goodness will appear in your gestures; and if you are bad-tempered this also will appear in your gestures—in short, everything that you do yourself passes over into the

> children and makes its way within them.... The inclinations that children develop depend on how you behave in their presence. (Steiner, 1998, p. 92)

> Every decision we make, every judgement we administer, every interaction with a child, demonstrates what we value and serves as a model for our students about what really counts in life, in being human. Every curriculum choice, every pedagogical method, every assessment, reveals what we consider important. All our decisions serve as influential models to children, whose dispositions, habits, and perspectives are formed under our care.... Our values are implicitly expressed in how we run our classrooms. They show in what subjects we teach, how we teach them, and in what we exhibit on the walls of the room. The children take all this in day after day, as unconsciously as breathing. It all works in subtle ways in the formation of their character. (Levy, 1996, p. 154-155)

> The way the teacher plays the mediator role conveys both an epistemology and an ethic to the student, both an approach to knowing and an approach to living. (Palmer, 1993, p. 29)

What the authors mentioned above, among others, imply in their thinking is that, if we want to be true educators, we need to work on our own development and on our relationship with the people, the knowledge, and the world around us so that we can be good role-models for our children and/or students. This is why, for example, Krishnamurti advises teachers to work on their own conditionings on a regular basis and why Steiner asks his teachers to meditate on a daily basis and systematically participate in workshops and staff meetings designed to promote their self-development.

While I had read about and talked to many other authors who discuss the importance of the quality of being of teachers, it was only after I developed this research that I came to fully understand the importance of this matter. The findings of my study point to the fact that, no matter how much we like the educational philosophy of the school, and how ideal the kind of structure offered by the school may seem for our child, if the teacher who is going to be with her doesn't have certain qualities, chances are that the school will not work well for her. In order to illustrate this point, I would like to cite the case of a student, Clara (9 years old), who I met during my fieldwork at the BWS. Before Clara studied at the BWS, she used to go to a local Waldorf school.[3] Her parents very much appreciated the Waldorf approach and thought it would be perfect for Clara. See for instance the following words by her mother, Deborah:

> I don't think there is an approach to education that more consistently daily touches all of those important pieces of learning for children from developing their fine motor skills to music, to art, all the handwork they do, as well as the story-telling, the spiritual piece, it is really children coming into who they are ... [Clara] is a Waldorf kid! She loves music, she loves arts, she is physical, she loves stories, she is really into the imaginative world....

While Clara's parents very much appreciated—and still appreciate—the Waldorf approach, they had to move her to the BWS:

> Clara went to first grade with this teacher who really scared her ... he hit her with a pointer one day, and punished students in different ways.... No matter how much we worked on it ... he was really glued on where he was and didn't get how to work with it.... So it really got to a crisis point and that's when we said we need to pull her out of this environment. She lost a lot of self-esteem last year, she was not a good learner, she would never do anything good enough.... (Deborah)

Clara had to leave the Waldorf School not because she or her family didn't like this school (her brother actually still studies there). The reason why Clara and some of her classmates left this school was that the teacher who they happened to have, "just couldn't live up to the school's educational philosophy," [4] as Clara's mother explained. Clara, who had almost completely lost her self-esteem and joy for learning while studying with this specific teacher, regained these qualities in her new school, the BWS. Her mother comments:

> It was an incredible process to see this little girl lose her sense of self with the other teacher [the Waldorf teacher] and now to see her with someone who she feels comfortable with. How much love to learn!... And how capable she is to take things up! *It wasn't really Waldorf that*

> *wasn't working for her, it was her relationship with her teacher which really didn't work....* I wish I had a video. It would be so great, a video of her last year in that learning environment, and her this year in this learning environment. (Emphasis mine)

Clara came to like the new school and started doing really well there—the BWS—not only because it was a good school, but because of the teacher who came to work with her, who was—and still is—a wonderful person and professional. As I fully described some of the qualities of this teacher, Cynthia Hendel, for now I would like to present some of the qualities which were common to all the teachers who seemed to be most effective in dealing with students in the three schools where I did my research. Among other things, they all:

- Invested time and energy in their personal and professional development.
- Systematically reflected on their practice. For example, if they had a problem with a student or group of students, they would often ask themselves questions such as: "What is really going on here?" "Am I causing this situation for some reason?" "Am I acting out of a commitment to students' learning and development or from my egoistic desire to control them?"
- Had a positive attitude towards life and people in general. They tended to see, for instance, the goodness of people and of the specific situations that happened with them. They also tended to celebrate the beauty of small things such as that of the rain poring, of a snowfall, or of a flower spurting.
- Enjoyed being with children and had a sense of gratitude for being able to work with them.
- Believed that all children wanted to learn and would learn regardless of their behavior and difficulties.
- Had a sense of admiration and respect for the uniqueness of each of their students.
- Taught from their "big Self" rather than from their "little self" or ego. As J. Miller (personal communication, 2000) once said, those who teach from the "big Self" or from their soul, teach from a

place of true love and not from selfish concerns, which helps them avoid unnecessary conflicts with students.

- Had a mindful presence in the classroom and lived in the moment with their students. In other words, when they were with their students, they were with their students as if there were nothing else more important for them to do.
- Dialogued with students and did their best to understand them and to address their specific needs.
- Had good leadership skills.
- Dedicated themselves to building healthy relationships and a sense of community among all those involved with the education of children (students, school staff, and families).
- Focused on the quality of the learning processes rather than just on the results of these processes.
- Acknowledged children's strengths and beauty and avoided reinforcing their weaknesses and bad behavior.
- Were sensitive and responsive to children's specific learning styles and rhythm.
- Were thoughtful, caring, kind, resourceful, and fair in terms of sharing their time and energy with children (gave attention to all students, and not only to the loudest, most problematic, and most verbal ones).

We all know developing these qualities is hard and demands long-term work. Therefore, those of us who are thinking about or already deal with the education of other human beings, need to urgently start dedicating ourselves to our self-development so that we can be good models for the people with whom we are working. And those of us who directly or indirectly deal with the education of prospective and/or continuing teachers need to start programs that can help them work on their qualities of being so they too can be good models for students.

Before ending this discussion, it is important to note that, not only should teachers have certain qualities of being in order to be effective as educators, but their way of being should be coherent with the educational philosophy of the school they work in. As we could see with the example of one of the SAU teachers, Greg, even though he was a wonderful person and great teacher, his way of thinking and living his life was not consistent with that of most members of the school community. Therefore, he felt he couldn't be truly himself at the school and could not fully implement his beliefs in his practice as a teacher. This obviously limited his work and relationship with many participants of the SAU community. Differently from Greg, Amy felt very much at home while in the SAU. After all, she had not only been raised and educated by two of the main founders of the school, but also did all her elementary education there. This contributed a great deal in the quality of her teaching and relationship with her students and their families.

As we choose a school for our child we need to consider not only factors such as school location and price. We need to be aware that different kinds of schools have different kinds of gifts and limitations. We also need to (1) examine our family values and whether these values are compatible with those of the school; (2) closely observe who our child is and what kind of structure he/she needs in order to optimize his/her learning and development; and (3) consider who the teacher who is going to be working with our child is. If he or she doesn't have certain qualities of being and is not personally ready to promote the school's goals, the school experience is likely not to be a positive one. Words about love, trust, respect, caring relationships, and appreciation for the beauty of the people and for the world around us are not enough. Teachers need to be ready to live what they believe in so that their attitudes will not challenge their own discourse and will not contradict the educational philosophy of the school.

Where Do We Go from Here? Implications for Further Research

More research on different kinds of educational institutions attempting to work within a holistic perspective is necessary. It would be especially enriching to investigate schools that work with the notion of holistic education as a way of educating people to achieve enlightenment, and freedom from conditionings and from any other sort of attachments, since there is still so little written about it in the Western world. Considering that no single school can meet the whole range of human needs, and different kinds of schools seem to work best for different kinds of families and students, it would also be important to investigate what kind of people students from different types of schools become. With information about the outcomes of schools, we would have more elements to discuss and make more informed decisions about which one seems to be most appropriate to our needs.

The experiences of different schools on their own individual journeys and in comparison with each other might teach us invaluable lessons about the complexities involved in the implementation of holistic ideas in education. As we develop more and more research in different kinds of schools that are trying to work from a holistic perspective and on their outcomes, we will be able to build a more solid basis of knowledge on which to construct more grounded and more truly holistic, holistic education theory and practice.

Notes

1. This was actually the case of the School Around Us: The school community spent so much time dealing with the social and emotional dimensions that the intellectual aspect was neglected. Ironically, the negligence in relationship to this aspect negatively affected students' emotional health: their self-esteem.
2. This school was not the same as the one where I did my research.
3. The same school where George goes.
4. It is important to note here that in Waldorf education, the teacher usually stays with the same group of students from first to eighth grade. The school where Clara used to go to, which was the only Waldorf School close to the town where she lived, offered only one class of each grade. This meant that if she and some of her classmates stayed in the school, they would probably be stuck with the same teacher whom they didn't like for eight years.

Bibliography

Berman, C. (1994a). *The School Around Us: 25 years.* Kennebunkport, ME: School Around Us Press.

Berman, C. (1994b). Video-taped alumni interview.

Beste Einsichten. (1981, December 14). *Der Spiegel*, pp. 70-73.

Board of Trustees of the School Around Us (1997). The School Around Us educational program handbook. Arundel, ME: School Around Us Press.

Bogdan, R., & Biklen, S. K. (1982). *Qualitative research for education: An introduction to theory and methods.* Boston: Allyn and Bacon.

Bordage, J. and Clark, C. (1989). Videotaped response to alumni questionnaire.

Bowers, C. A. (1993). *Education, cultural myths, and the ecological crisis: Toward deep changes.* Albany, NY: State University of New York Press.

Coles, R. (1993). *The call of service.* New York: Houghton Mifflin.

de Souza, D. (2000). "Holistic education: Learning from the experiences of three holistic teachers." Unpublished doctoral dissertation. Harvard University, Cambridge, MA.

Dewey, J. (1963). *Experience and education.* New York: Macmillan.

Dewey, J. (1964). *John Dewey on education.* New York: The Modern Library.

Dewey, J. (1966). *Democracy and education.* New York: The Free Press.

Dwight, G., Honnet, E. P. & Migliore, B. (Eds.) (1991). *Research agenda for combining service and learning in the 1990s.* National Society for Internships and Experiential Education.

Edmunds, L. F. (1992). *Rudolf Steiner education: The Waldorf School* (3rd ed.). London: Rudolf Steiner Press.

Forbes, S. (1999). "Holistic education: an analysis of its intellectual precedents and nature." Unpublished doctoral dissertation, Oxford University, U.K.

Gallegos, N. R. (1999). *Educación holista: pedagogía del amor universal.* México, DF: Paz.

Gallegos, N. R. (2000). *El Espíritu de lá educación: Integridad y transcendencia en educación holista.* Guadalajara: Fundación Internacional para lá Educación Holista.

Greenberg, D.; Sadofsky, M. (1992). *Legacy of trust: Life after the Sudbury Valley school experience.* Framingham, MA: Sudbury Valley Press.

Hammersley, M. & Atkinson, P. (1995). *Ethnography: Principles in practice* (2nd ed.). London: Routledge.

Harwood, A.C. (1992). *The recovery of man in childhood.* New York: Myrin Institue of New York.

Hedin, D. P. (1989). "The Power of Community Service" *Proceedings of the Academy of Political Science,* 37 (2), 201-213.

Henry, M. E. (1993). *School cultures: Universes of meaning in private schools.* Norwood, NJ: Ablex Publishing Corporation.

Holland, W. B. (1981). *The Waldorf Schools: 32 questions and answers.* Inverness, CA: Children's Book Service.

Holt, J. (1972). *Freedom and beyond.* New York: E.P. Dutton & CO.

Holt, J. (1982). *How children fail.* New York: Delta/Seymour Lawrence Edition.

Holt, J. (1983). *How children learn.* New York: Delacorte Press/Seymour Lawrence.

Kessler, R. (2000). *The soul of education: Helping students find connection, compassion, and character at school.* Alexandria, VA: ASCD.

Krishnamurti (1981). *Education and the significance of life.* San Francisco, CA: Harper & Row.

Leeichter, Hope Jensen. (1980). "A note on time and education." *Teachers college Record,* 81(3).

Levy, S. (1996). *Starting from scratch: One classroom builds its own curriculum.* Portsmouth, NH: Heinemann.

Lincon, Y. S. & Guba, E. G. (1985). *Naturalistic inquiry.* Beverly Hills: Sage.

Marshak, D. (1997). *The common vision: Parenting and educating for wholeness.* New York: Peter Lang.

Maxwell, J. (1996). *Qualitative research design: An interactive approach.* Thousand Oaks, CA: Sage.

Mello, G. N. (1982, February) *Pesquisas em Educação: Questões teóricas e questões de método. Cadernos de Pesquisa*[Research in education: Theoretical questions and questions of method]. Fundação Carlos Chagas, *40.*

Miles, M., & Huberman, M. (1994). *Qualitative data analysis* (2nd ed.). Thousand Oaks, CA: Sage.

Miller, J. (1988). *The holistic curriculum.* Toronto: OISE Press.

Miller, J. (1993). *The holistic teacher.* Toronto: Ontario Institute for Studies in Education Press.

Miller, J. (1994). *The contemplative practitioner: Meditation in education and the professions.* Westport, CT: Bergin & Garvey / Greenwood.

Miller, J. (2000). *Education and the soul: Toward a spiritual curriculum.* Albany, NY: State University of New York Press.

Miller, R. (1993). "Holistic education in the United States: A 'new paradigm' or a cultural struggle?" *Holistic Education Review, 6* (4), 12-19.

Miller, R. (1995). "Freedom in a holistic context." *Holistic Education Review, 8* (3), 4-11.

Miller, R. (1997). *What are schools for? Holistic Education in American Culture* (3rd ed.). Brandon, VT: Holistic Education Press.

Montessori, M. (1964). *The Montessori method* (11th ed.). (A.E. George, Trans.). New York: Schocken Books.

Montessori, M. (1966). *The secret of childhood* (M.J. Costelloe, Trans.). Notre Dame, IN: Fides Publishers.

Montessori, M. (1967). *Spontaneous activity in education* (4th ed.). (F. Simmonds, Trans.). New York: Schocken Books.

Nakagawa, Y. (2000). *Education for awakening: An eastern approach to holistic education.* Brandon, VT: Foundation for Educational Renewal.

Neill, A. S. (1960). *Summerhill: A radical approach to child rearing.* New York: Hart.

Neill, A. S. (1966). *Freedom—Not license!* New York: Hart.

O'Neil J. (1993). "Learning through service." *Update. Association for Supervision and Curriculum Development, 35*(6), 5-8.

Orr, D. (1994). *Earth in mind: On education, environment, and the human prospect.* Washington, DC: Island Press.

Palmer, P. J. (1993). *To Know as we are known: Education as a spiritual journey.* San Francisco: Harper Collins.

Palmer, P. J. (1998). *The courage to teach.* San Francisco: Jossey-Bass.

Patton, M. Q. (1990). *Qualitative evaluation and research methods* (2nd ed.). Newbury Park, CA: Sage.

Reinsmith, W. A. (1989). "The whole in every part: Steiner and Waldorf schooling." *The Educational Forum, 54* (1), 79-91.

Rousseau, Jean-Jacques. (1957). *Émile* (B. Foxley, Trans.). New York: E. P. Dutton.

School Around Us (1970). School brochure.

Saviani, D. (1984). *Escola e Democracia; teorias da educação; curvatura da vara; onze teses sobre educação e política* [School and democracy; educational theory; stick bending; eleven theses about education and politics. (3rd ed.). São Paulo: Cortez e Autores Associados.

Silcox, H. (1993). *A how to guide to reflection: Adding cognitive learning to community service programs.* Philadelphia: Brighton Press.

Smith, G. A. (1992). *Education and the environment: Learning to live with limits.* Albany, NY: State University of New York Press.

Stanton, T. (1990). "Service Learning: Groping toward a definition." In *Combining service and learning* (pp. 65-67). National Society for Internships and Experiential Education.

Steiner. R. (1976). *Knowledge of the higher world and its attainment.* London: Rudolf Steiner Press.

Steiner, R. (1995). *The spirit of the Waldorf school.* (R. F. Lathe & N. P. Whittaker, Trans.). Hudson, NY: Anthroposophic Press.

Steiner, R. (1998). "The kingdom of childhood." In R. Trostli (Ed.). *Rhythms of Learning: Selected Lectures by Rudolf Steiner.* Hudson, NY: Anthroposophic Press.

Steiner, R. (1999). *The philosophy of freedom* (M. Wilson, Trans.). London: Rudolf Steiner Press.

Swanson, E. (1994). "Incorporating service learning into writing and literature classes." In *Building community: Service learning in the academic disciplines* (pp. 131-137). Denver, CO: Colorado Campus Compact.

Trostli, R. (1991). "Educating as an art: The Waldorf approach." In R. Miller (Ed.), *New directions in education: Selections from Holistic Education Review* (pp. 338-351). Brandon, VT: Holistic Education Press.

Trostli, R. (Ed.). (1998). *Rhythms of learning: What Waldorf education offers children parents & teachers*. Selected lectures by Rudolf Steiner. Hudson, NY: Anthroposophic Press.

Yoshida, Atsuhiko (October, 1999). "Interface of holistic changes in Japanese schools and the Waldorf education." Paper presented at the International Conference on Holistic Learning: Breaking New Ground. Toronto.

Yoshida, Atsuhiko (November, 2000). "Explorando la educación holista en la propria cultura." Paper presented at the VIII Conferencia Internacional sobre educacion holista. Guadalajara, Mexico.

Appendix

Research Methodology

The Research Questions

The three case studies revolved around two questions:

1. How do the schools express their philosophy of education and conceptualize their practice?

2. How do they implement their educational philosophy and balance freedom and structure in the curriculum?

Data Collection

I collected the data for the case study on the School Around Us (SAU) from March to June of 1997, two to four times a week. The data for both the Sunshine Waldorf School (SWS) and the Bellwether School (BWS) were collected from January to May of 2000. In the case of the SWS, which was located close to where I lived, I visited the school two to four times a week except for the weeks I was away to study the BWS. In the case of the BWS, which was located in a different state, I observed the school for periods of four to five days every few weeks. I spent the whole school day at each of the schools whenever I visited them, and often went earlier and stayed later than the regular school hours to interview teachers, students, and parents, as well as to review school materials and students' work. After I finished my fieldwork, I conducted a few telephone interviews with some teachers so that I could follow up on some questions that needed more investigation. In the case of the SAU, I conducted a total of 23 observations in the classroom, 47 audio-taped interviews, and some non-audio-taped interviews. In the case of BWS, I completed 15 full days of observation in the classroom, 37 audio-taped interviews, and some non-audio-taped interviews. In the case of the SWS, I conducted 23 full days of observations in the classroom, 26 audio-taped interviews, and many non-audio-taped interviews (I was asked not to audio-tape my interviews with the SWS students). Besides the full day observations in all the schools, I also observed school events such as field trips, festivals, plays, teacher/parents meetings, seminars, and workshops. I used three basic strategies to collect data: observations, interviews, and document review.

Observations

The main goal of my observations was to understand how the teachers actually implemented the school's educational philosophy, and how they dealt with the issue of freedom and structure. The observations also generated data, which helped me investigate and understand the schools' philosophy of education and conceptualizations of their practice.

As the School Around Us was small and had only three teachers who worked with the same group of students, I shared my time equally observing and interviewing each of them. As both the BWS and the SWS were larger, while I did observe different classrooms and interviewed various teachers, I focused most observations and interviews on one teacher of each of these schools so that I could collect in-depth data.

I audio-taped many of the classes and meetings both at the SAU and at the BWS. I did not audio-tape classes at the SWS since this procedure would feel intrusive in that environment. I took extensive fieldnotes in the three schools and used my laptop most of the time in order to take notes. I would confer my field notes every night after the field research so that I could fill in with information I hadn't had the time to write down during observations. Before I did my final data analysis, I transcribed pieces of the audio-taped classes and school meetings that were relevant to answer my research questions, so that I would not miss important details of what happened during these classes and meetings.

INTERVIEWS

I interviewed teachers, students, some parents, board members (when the school had board members) and a few ex-students of each of the schools. I audio-taped most of these interviews and transcribed all of them. When it was not possible to audio-tape, I took extensive notes on all details of the interviews. In the case of the School Around Us, I audio-taped and transcribed twenty interviews with the teachers, sixteen interviews with the students, eight interviews with parents, two interviews with two ex-students, and one interview with one of the founders of the school. In the case of the BWS, I audio-taped and transcribed 14 interviews with teachers, 13 with students, 7 with parents, 3 with one of the founders and member of the school board. In the case of the SWS, I conducted 22 interviews with teachers, 2 with parents, and 17 non-audio-taped interviews with students, one with an ex-student and one with a teacher at a charter Waldorf School.

My first interview with most teachers was semi-structured (Bogdan & Biklen, 1982). During this interview, I asked them questions about the school's philosophy of education, personal educational philosophy, teaching background, conceptualizations of their teaching practice, and difficulties they faced as they tried to implement their conceptualizations. This interview gave me general background knowledge of the main issues I wanted to study. After that, I conducted many "informal conversational interviews" (Patton, 1990, p. 281) in which I asked them (1) follow-up questions to my first interview; (2) questions related to what I observed in their classes (e.g., why they developed certain activities, made certain comments, used certain materials, went on specific field trips); (3) questions related to the difficulties they faced as they tried to implement their ideas (e.g., why the students were so resistant to participate in a certain activity; how the implementation of their philosophy of education was affected by the way the school was run and organized and by the kinds of resources available to them; and (4) questions about their understanding about which degree of freedom and structure was offered by the school and by themselves in their classroom.

I conducted numerous informal conversational interviews with the students so that I could have their perspectives in relationship to the school's and the teachers' educational philosophy and practice. During these interviews, I asked them about such things as their educational background, why they were studying where they were studying, the things they liked about the school, the things they would like to see changed in the school, why they behaved in specific ways during the school meetings and/or classes, and what they had learned in specific school experiences. Because the student-teacher ratio at both the SAU and the BWS was low and the schedule of the school allowed for long breaks between each class, I was able to amass substantial data by conducting informal conversational interviews with the teachers and students on most of the occasions I visited the school. In the case of the SWS, I wasn't able to interview as many students with the same level of depth as I did in the other two schools since the schedule was tighter (they had less free time) and since the teachers themselves asked me not to talk with students about certain subjects. From their perspective, children should not be called to think about certain issues and make certain judgment calls before a certain age.

My interviews with the parents revolved around three main questions: Why they chose that specific school for their children, what were the things they liked the most about the school, and what kinds of things they would like to see changed in the school, if they thought there was anything to be changed. My interviews with the ex-students explored their perceptions on the main strengths and weaknesses of the schools. The data from these interviews offered me contextual information on the school, the school community, and the school community's perspectives on the school. This helped me to better understand and further explore not only the educational philosophy and practices of the school and of the teachers, but also the main problems that affected the latter's practice.

Document Review

In order to further understand the educational philosophy and practice of the three teachers, I watched videos and reviewed books on the schools. I read, for instance, a book on the SAU by Claudia Berman (1994a), and various books on Waldorf education by Rudolf Steiner and some of his followers. I also reviewed other written materials related to the schools such as parents' handbooks, handouts teachers prepared for the students, and samples of students' work.

Data Analysis

The data analysis went hand in hand with the data collection. After each of the interviews and observations, I typed up my field notes, listened to the audio-taped classes and interviews, and wrote up analytic notes with insights and questions for further observation and interviews (Miles and Huberman, 1994). Every few days I also wrote memos in which I identified important themes, patterns and contradictions in the data collected (Hammersley & Atkinson, 1995) to both organize and synthesize my findings and to further focus the data collection.

I analyzed the data in two stages. First I analyzed the data related to the educational philosophy and practice of each of the schools and wrote up separate case studies. Second, I developed a cross-analysis of the data related to each of the schools, examining similarities and contrasts between their educational philosophy, practice, and understandings on the issue of freedom and structure.

I based my preliminary data analysis of the interviews, observations and document review on a sensitizing framework and on analytical categories which were developed with basis on the literature on holistic education, interviews with major holistic education scholars, and pilot observations in schools that defined themselves as holistic. But, I did modify this framework many times according to the conceptual categories that grew out of the data. As I noticed some categories didn't seem relevant as other themes that emerged from the data, I eliminated them and added new categories to the analytical framework.

Ethical Concerns

Before I started my research, I talked with teachers from each of the schools, explained the goals of my research, and asked their written consent to both observe their classes and interview them. I also sent informed consent forms for the parents of all the students who I was planing to observe and interview and asked for their written permission so that I could do so. Both in the case of the SAU and BWS, I also sent informed consent forms for the children to fill in and sign. I did not do so in the case of the SWS, since the teachers thought children should not be called to make this kind of decision on their own.

I also asked the teachers who were involved in the study, in writing, whether or not I should use their real names and the real names of the school in any publications that resulted from my study. Both the SAU teachers and the BWS teachers agreed that I should use their real names, which I did. I had to change, however, the name of the Waldorf School and the names of all the teachers who worked there. I used pseudonyms for all children and parents from the three schools, since a few of them preferred not to be identified.

When I first met the teachers and parents, I told them I was aware that my research could have some kind of impact on the participants, and that I would like to include their reactions to my findings in my final report, so they could speak for themselves as I spoke about them. As soon as I finished drafts of my work, I sent these drafts to the teachers who were willing to read them, and incorporated their comments in the manuscript.

Trustworthiness

Internal validity

The Problem of Bias. I started my research with some preconceptions, which came primarily from two sources: from my traditional educational background, and from my studies on the holistic education movement as it is expressed in the literature. My traditional educational background—one that tended to emphasize intellectual development over other aspects of our development—led me to expect, for instance, to see the SAU teachers and students focusing more on academic work than they actually did. The theory of holistic education influenced me in the sense that I expected to observe the teachers emphasizing and trying to implement all the main themes of the holistic education movement into their practice. These two sources of preconceptions were, in some ways, paradoxical. While the first focused on the importance of promoting the intellectual self of the child, the second focused on the importance of looking at the child as a whole person and of promoting the balanced development of all aspects of her being. The antithetical aspects of these tendencies helped me balance my biases, which supported me as I tried to stay aware and honest about them while I collected and analyzed the data, and wrote my research report and this book.

Another major source of bias for me was my commitment to share my findings with the school community, and my desire not to offend or hurt anyone's feelings with comments that had potentially negative implications about their educational philosophy and practice. I tried to remain aware of this problem throughout the whole research process. While I was careful in the way I worded my interview questions that investigated areas of weaknesses, I tried not to let this affect my commitment to do probing research. And while I was careful in the way I chose to express my findings, I tried not to let this compromise the content of what had to be said.

The Problem of Reactivity. This problem refers to my influence on the setting and on the individuals studied. As I will further explain, the three schools included in my study were open to parental participation in the classroom. In the case of the SAU and of the BWS, they were also quite open to, and used to visitors. Because the teachers and students from the three schools were used to interacting with different people in the classroom, my presence there probably did not affect their activities and behavior as much as it would probably affect those of people in other educational settings.

One factor that did help me decrease the problem of reactivity was the fact that I spent a lot of time in the schools and participated in many events promoted by them. This helped the communities of each of the schools get more used to and feel more comfortable with my presence. See for instances the following quotes—one by Greg and one by Amy, from the SAU—when I asked the first about my influence on his practice, and the second about my influence on students' behavior:

> You saw plenty of frustration, you saw day to day reality, times that were totally ineffective ... other times being successful. You saw all, I don't think you have influenced so much and you asked good questions.... For me you had a pretty accurate view of what was going on.... You saw my best and my worst and the inherent problems that I have in terms of inconsistencies. I didn't change anything for you, I felt

> completely comfortable, and not at all under a microscope, so I felt free to do my best and my worst. (Greg, audio-taped telephone interview, 7/1/98)

> They pretty much behaved the way they did before you.... I think that you got a good idea of how the school runs because I think that you were there enough that they got used to you and they didn't feel like they had to hide anything and they could show their true colors. (Amy, audio-taped telephone interview, 8/20/98)

The fact the teachers claimed that I didn't affect their practice, at least in significant ways, doesn't necessarily mean that I in fact did not affect them in subtle ways. Therefore, as I analyzed the data, I tried to keep in mind the ways in which I might have influenced what the participants in my research did and said. For instance, when I analyzed interview data, I took into account the context in which individual interviews were conducted, and how this context and the ways in which I asked the questions might have influenced what the participants said during these interviews.

Descriptive, interpretive and theoretical validity

In order to attend to the descriptive, interpretive and theoretical validity (Maxwell, 1996) of my research, I attended to the following strategies: First, I tape recorded most of my interviews and observations in addition to taking detailed field notes. In addition, I made verbatim transcriptions of all audio-taped interviews (I transcribed the entire interviews) and of parts of classes and school meetings which were relevant to answer my research questions, so that the data for description and analysis would be complete and accurate.

Second, I used triangulation of data through sources and methods (Lincoln & Guba, 1985). In other words, I observed classes, school meetings and teacher and student interactions; I interviewed teachers, students and parents; and I reviewed handouts, students' work, and school documents (e.g., the school handbook and videos). These different sources and methods allowed me to consider and contrast different perspectives on the issues under consideration, to check the accuracy of the data I was gathering, and to probe my ongoing interpretations.

Third, as I collected and analyzed the data, I tried to make sense of how the participants in my research understood what was going on instead of imposing my own interpretative framework on them. While I started my research with a sensitizing framework to guide my data collection and analysis, I modified this framework according to the most important themes that emerged from my data. As I worked on the analysis of these themes, I stayed as close as possible to their own language, metaphors, and emic concepts. For example, a common expression that the SAU teachers and students used in the school was "to do the consequences," which was something the students were sometimes asked to do after badly misbehaving three times in a same class. In the entire time I observed the school, they invariably used the word "consequence" instead of "punishment." Curious about this choice, I asked one of the teachers, Kendall, about it. She explained that for the school community, the word "punishment" implies "paying back" for something bad you did. The word "consequence," on the other hand, implies that everyone's acts always have implications, and therefore everyone should be careful when he/she chooses to behave in certain ways. As I became aware of the conscious choice that the SAU made to use the term "consequence" instead of "punishment," whenever I had to think about and refer to what happened when students misbehaved, I used the expression "consequence," taking into consideration the concept behind it for the members of the SAU community.

Fourth, as I have previously mentioned, my data analysis went hand in hand with my data collection. This allowed me to constantly probe my emerging theories of what I was observing in the field in two ways: (1) I looked for confirming and discrepant evidence to support or to modify my interpretations of the data (Maxwell, 1996); and (2) I conducted many "member checks" (Lincoln & Guba, 1985, p. 314-316). In other words, I checked the accuracy of my ongoing understanding of the teachers' philosophy and practice through many conversational interviews with the teachers.

Fifth, at the end of my research, I shared a draft of my final report with some of the participants of my study for a final "member check" (Lincoln

& Guba, 1985, p. 314-316). Amy, Greg, and Kendall (the three teachers from the School Around Us), Lauren from the SWS, and Cynthia from the BWS, the teachers who I cited the most throughout my work (I spent most of my time in their classrooms), read my final reports on their school and endorsed the validity of my study.

External validity or transferability

As I conducted an in-depth qualitative inquiry, I cannot establish the external validity of my study. However, I do "provide the *data base* that makes transferability judgments possible on the part of potential appliers" (Lincoln & Guba, 1985, p. 316). In other words, I provide rich contextual descriptions so that "anyone else interested in transferability has a base of information appropriate to the judgment" (Lincoln & Guba, 1985, pp. 124-125). Instead of generalizing my findings, I attempted to generate theory that may help others interested in alternative and holistic approaches to education to further elaborate on the theory and practice of this kind of these approaches.

Chapter 11

Questions and Issues

Accountability Issues

Waldorf students usually do not take academic tests until the later grades. In the case of the SWS, they start taking tests only in 6th grade as a way of getting them prepared for the standardized tests they are likely to have to take when they leave the school. Since they are not tested before then, an accurate view of their developmental and learning level depends on the teacher's observational and organizational skills.

The fact that Waldorf teachers stay with the same group of students for a long period of time has the advantage that it allows them to know their students well and gives them time to consistently work with individual children according to their personal needs. On the other hand though, in cases in which the teachers are not able to detect a learning difficulty in the child, the latter is likely to go a long way before somebody else notices the problem. The SWS is trying to prevent this problem by having a routine evaluation with all first and second grade students in aspects such as learning differences, learning styles, dominance, reading skills, motor coordination, memory skills, ability to follow directions, and behavioral problems. The specialist who does this evaluation with the children gives teachers a good picture of where each child is in terms of learning and development, and when necessary, refers her to some kind of "remedial work" or special tutoring. He/she also helps the teacher devise strategies to deal with the needs of individual children.

Provocative Books on Teaching and Learning that Fosters Personal and Cultural Transformation From the Foundation for Educational Renewal

Foundations of Holistic Education Series

Caring for New Life: Essays on Holistic Education by Ron Miller (2000)

Education for Awakening: An Eastern Approach to Holistic Education by Yoshiharu Nakagawa (2000)

"Under the Tough Old Stars": Ecopedagogical Essays by David Jardine (2000)

Unfolding Bodymind: Exploring Possibility Through Education. Edited by Brent Hocking, Johnna Haskell, and Warren Linds (2001)

Holistic Education: Pedagogy of Universal Love by Ramon Gallegos Nava; translated from the Spanish by Madeline Newman Rios and Gregory S. Miller (2001)

Nurturing Our Wholeness: Perspectives on Spirituality in Education. Edited by John P. Miller and Yoshiharu Nakagawa (2001)

The Primal, the Modern, and the Vital Center: A Theory of Balanced Culture in a Living Place by Donald W. Oliver, Julie Caniff, and Jouni Korhonen (2002)

Holistic Education: An Analysis of its Nature and Ideas by Scott Forbes (2003)

Other Titles of Interest

A Free Range Childhood: Self-Regulation at Summerhill School by Matthew Appleton (2000)

Creating Learning Communities: Models, Resources, and New Ways of Thinking About Teaching and Learning. Edited by Ron Miller (2000)

For more information,
Visit the Foundation on the Web at
<www.great-ideas.org> or
Phone 1-800-639-4122.